AF327263

THE CULTIVATED HEMLOCKS

Tsuga canadensis 'Sargentii'. U.S. National Arboretum, Washington, D.C.

THE CULTIVATED HEMLOCKS

John C. Swartley, Ph.D.

Revised by Humphrey J. Welch

Technical Editor: Theodore R. Dudley, Ph.D.
Research Botanist, U.S. National Arboretum,
U.S. Department of Agriculture

Biographical essay about the author
by Robert L. Fincham

TIMBER PRESS
Portland, Oregon
1984

© Timber Press 1984

All rights reserved.
Printed in the United States of America.

ISBN 0-917304-74-8

TIMBER PRESS
P.O. Box 1631
Beaverton, Oregon 97075

CONTENTS

FOREWORD

If ever a longsuffering public has had to exercise patience it will have been over this book.

It was as far back as 1939 that Dr. A. B. Stout, writing in the Journal of the New York Botanical Garden, announced that a comprehensive survey of *Tsuga* forms was in preparation by a certain John C. Swartley. This referred to a Thesis with the title "Canadian Hemlock and its Variations" then being prepared by the present author for submission to Cornell University the same year, for his hoped-for degree as Master of Science. Unfortunately, although it achieved its object this thesis was not published. It is clearly quite impossible for every Thesis by every student at every university to be published, but had the authorities at Cornell realised that this particular Thesis would remain the only material study in this genus for forty years, and that they would be asked for copies of it by foreign governments and botanical gardens at home and abroad they might have decided otherwise. Attempts were made to persuade the author himself to do something about it. In 1945, at the request of the late Peter den Ouden who was then collecting material for the work that was eventually to appear as the MANUAL OF ORNAMENTAL CONIFERS (Den Ouden and Boom. 1965), he travelled over the ground covered in the original Thesis and prepared a Supplement. Although this was used in the book just mentioned it was not itself published, and it was not until about 1970 when I was able to add my voice to the many that were clamouring for a "Swartley on Tsuga" book to appear that the author bowed to the clamour and gave it his attention. The manuscript was soon in the hands of a publisher, but by the time I was drawn into the matter, ten years or so later, there was still no book, and by then the manuscript had become so out of date as to need a careful revision.

In the meantime, alas, disaster had struck in the form of a breakdown in Swartley's (now Dr. Swartley's) health, gradually but progressively curtailing his ability to get around botanising, and also affecting his powers of concentration. Dr. T. R. Dudley, Research Botanist at the United States National Arboretum, Washington D.C., had been brought in as Technical Editor and Adviser, in which capacity he clearly did a great deal of work, but it must have been done with rather a feeling of fighting a losing battle against time and the author's failing health.

Also in the meantime changes in my own life had been occurring. My own book MANUAL OF DWARF CONIFERS had appeared (and clearly the way it dealt with certain nomenclatural problems, especially in the use of *Groups* could also be used in a monograph on the hemlocks), and the Royal Horticultural Society, who are the International Registration Authority for Ornamental Conifers, hearing of my retirement from business had roped me in as Registrar.

It was clear that the work I should have to do on the hemlocks in preparing the draft Register would run parallel with the review that had become necessary on the Swartley manuscript. One is often unaware of the forces at work as a situation develops and be unable to remember how it all happened, but quite unexpectedly I found myself having agreed to up-date the monograph, incorporating some of the ideas regarding layout, etc., from my own book and making sure that nothing would be at conflict with the work being done on the Conifer Register in preparation.

It has been an interesting task, rendered difficult from the inaccessibility of the author's own wealth of knowledge and by the passing away of several collectors and experts such as Henry J. Hohman and Fred Bergman. And, of course, working in England on a book mainly devoted to cultivars selected in and almost exclusively grown in America, is far from the best arrangement. Dr. Dudley has continued his efforts and we all owe him a deep debt of gratitude. In addition, I am personally most grateful to those growers and collectors who have responded to my appeals for help over the newer American cultivars not known to me. Special mention must here be made of Layne Ziegenfuss, Robert Fincham, John Mitsch, Jean Iseli and Richard Bush.

The usual proviso made when a work has been revised by someone other than the author himself (i.e. that he retains all the credit and the reviser carries the blame for all the shortcomings of the revision) must on this occasion be taken seriously. Even the re-arrangement of existing material introduces plenty of scope for errors to creep in, and of course I am solely responsible for all new matter brought in. I can only hope not to get into too much trouble and disgrace.

Finally, great credit is due to our publisher for his enthusiasm and help, firstly in persuading his typesetters to work from a manuscript that fell way short of the "clean copy" that these craftsmen like to handle and secondly in the short time he has taken in turning dream into reality. It is never possible in life to make up for lost time, but Mr. Richard Abel of Timber Press must have been determined that no complaint of undue delay could be laid at his door.

HUMPHREY J. WELCH

May, 1983

AUTHOR'S ACKNOWLEDGMENTS

After a lifetime of interest in the cultivated *Tsuga* it would be impossible to remember, let alone name all the generous people who have given of their time and effort in helping me to bring this book to fruition. There are some, however, whose willingness to help has gone well beyond the call of duty.

Mr. Charles F. Jenkins was the first to encourage my study of hemlocks at Cornell University. His collection of hemlocks at his home, Far Country, Kitchen's Lane, Mt. Airy, Philadelphia, Pennsylvania, and the invaluable *Hemlock Arboretum Bulletins* which he published formed the starting point of my research. For his friendship, kindness, generosity, interest and encouragement over the years I owe much.

Mr. Jenkins died July 2, 1951, after several weeks in the Pennsylvania Hospital. His grave is in the Friends' burial ground at Upper Dublin, Montgomery County, Pennsylvania; it is shaded by one of his beloved hemlocks planted years before by his own hand. And so we lost a friend, a philosopher, writer, poet, historian and horticulturist. It was a privilege to have been associated with this unusual man of such varied interests and wide culture. (A portrait and a short biographical note will be found on p. 67).

For their encouragement and help in editorial work and details of descriptions, and in making their collection of hemlocks at Raraflora, Feasterville, Pennsylvania, available for research, I am deeply indebted and grateful to the late Fred Bergman and his wife Helene. Since his passing, alas, that remarkable collection has been broken up.

My visits to see Mr. Alfred Fordham, then Propagator at the Arnold Arboretum of Harvard University, Jamaica Plain, Massachusetts, were most rewarding. He was most gracious and helpful in having his assistant list all accession numbers of hemlock cultivars in the collection there. Mr. Fordham helped to locate, and called my attention to, many plants that would otherwise have been overlooked.

My thanks also go to Mr. Frank Stearns, then (1938-42) Chief of the Plant Protection Division, U.S. Department of Agriculture, Washington, D.C. for help in rating the relative importance of hemlock insect pests throughout the United States. I also consulted Mr. Robert F. Doren, formerly of the National Arboretum.

Whilst engaged in research for the Hemlock Thesis, I was privileged to spend some time with Liberty Hyde Bailey in the Bailey Hortorium and with Alfred Rehder in the Herbarium of the Arnold Arboretum. Each of these great men interrupted a busy schedule to discuss the scope of the proposed work and to offer suggestions and advice to a young entrant into their world of plants.

It is only right to put in a good word for the many nurserymen whom I visited and who gave generously of their time. Others gave valuable information by correspondence.

And thanks to my wife Elizabeth both for badgering me into finishing the manuscript and for typing and retyping it so many times.

For the indefatigable help of Dr. Theodore Robert Dudley, Research Botanist at the United States National Arboretum, Washington, D.C., for his constructive criticism and close editing of the manuscript, my gratitude is beyond words.

And, finally, my thanks go to Humphrey J. Welch for kindly undertaking to go through my book, adding fresh material necessary to bring it up to date, re-arranging the material where he considers (in the light of his experience with his own recent book) that the layout could be improved and—last, but by no means least—making sure that all would be in consonance with his other work as International Conifer Registrar.

JOHN C. SWARTLEY PH.D.
Chairman (1957-1966), Department of Horticulture,
Temple University
Ambler, Pennsylvania

May, 1983

ABOUT THE AUTHOR

Wanted a D.Sc.

Is there not some young man, (or woman), who has graduated from college in botany, horticulture, biology, ecology, dendrology, landscape gardening or some kindred science who would like to work on a thesis for a doctor's degree on the "Mutations of Tsuga Canadensis"? It would be a useful, worthwhile subject, a real contribution to scientific knowledge and fortunately there is now plenty of laboratory material.

If some graduate turns up who would be keenly interested and qualified for such an investigation, the Hemlock Arboretum would be glad to entertain him (or her) with the microscopes and other paraphernalia for as long as it might be necessary to identify and classify the specimens growing there and in addition make a grant towards the publication of the thesis when the work is completed.

This advertisement was printed by Charles F. Jenkins in his bulletin, *The Hemlock Arboretum at Far Country,* #19, July 1, 1937. In his April 1, 1938 bulletin, Jenkins announced the selection of a young man to perform this work at Cornell University. The young man's name was, of course, John C. Swartley. As from small acorns large oak trees grow, so through a research project begun at Germantown, Pennsylvania, a monumental work results and an obscure teacher-plantsman attains the fame he now so justly deserves.

John Clymer Swartley, the youngest of six children, was born on May 8, 1909 in Chalfont, Pennsylvania. He grew up on his parents' farm where he was expected to work very hard. As a youngster John earned spending money by picking potatoes for his neighbors. Something happened to John during his teens that would affect his physical abilities for the rest of his life. At sixteen, he contracted polio, then a little known disease. His right side has always been somewhat impaired as a result.

John was neither a hunter nor a fisherman, but he loved the out-of-doors and spent many days of his youth hiking. His love of nature led him to study the plant world. Although his formal education was in mathematics, his devotion to botany and horticulture soon became his life's work. After graduating from Doylestown High School, John attended the University of Pennsylvania, where in 1930, he was awarded a Bachelor of Science Degree in Mathematics Education. Now he would be able to serve in the noblest profession—education.

As often happens, reality brings a rude awakening. The majority of high school students need a firm hand to foster learning. John did not relish the constant pressure of the classroom situation. His first high school teaching position at East Mauch Chunk, Pennsylvania lasted until February, 1931. John never left education, but he taught in a high school situation only once more in his life.

After leaving his position, John did part-time horticultural work, including a term of graduate work in botany at the University of Pennsylvania during the spring term of 1932. He was employed as a propagator at the Morris Arboretum from May, 1933 until February, 1938.

He commuted to the arboretum from his aunt's home in Eureka in a Model A Ford. One morning after a severe snow storm, he was involved in an accident with a stolen truck. His Model A was demolished, and he spent some time in the hospital.

After leaving the hospital, John searched for housing close to the arboretum since he had no car and was walking on crutches. He obtained lodging at the home of Mr. and Mrs. William Parker who had two daughters. John courted one of the young ladies, and in 1937 John and Elizabeth Parker were married.

Elizabeth, a trained secretary, has typed many of John's papers and books throughout their married life. Even more importantly she provided financial support while John was earning his graduate degrees.

While working at the Morris Arboretum, John read Jenkins' advertisement in *The Hemlock Arboretum Bulletin*. Jenkins financed John's work on his monumental thesis, *Canada Hemlock and Its Variations*. The thesis was submitted for his Master's Degree from Cornell University in 1939. Unfortunately, copies of the thesis are in the possession of very few people. Jenkins had hoped to see the thesis published by the university, but it was placed in Cornell's library and never published.

After graduating from Cornell, John was awarded a scholarship by the American Paint and Chemical Company (AMCHEM) to Ohio State University. His dissertation on *Adventitious Root Initiation in Forsythia Suspensa* earned him his Ph.D. in 1942.

John tried teaching again, in the Cleveland public school system. After a very short time, he resigned and took employment as a landscape foreman for a large nursery. In 1943, he left the nursery and he and Elizabeth worked in war production at Thompson Products in Cleveland. They worked together on the 3:00–11:00 p.m. shift. John worked in the heat treating department making valves, and Elizabeth did secretarial work. John preferred this shift because it allowed him to maintain a vegetable garden outside of the city. Every morning he rode a bicycle from their apartment to his "victory garden".

In 1944 John and Elizabeth returned to her parents' home in Erdenheim, Pennsylvania, where he was hired as assistant groundskeeper at St. Thomas Church in Whitemarsh. While working at this job, John updated his thesis and sent the revised version to den Ouden and Boom to be used in their text, *Manual of Cultivated Conifers*.

From 1951 to 1966 John was Director of the Awbury Arboretum, at one time a private estate of fifty-seven acres in Germantown, Pennsylvania, established in 1852 and bequeathed to the City of Philadelphia in 1917.

In 1952 in Ambler, Pennsylvania, the Swartleys opened the Holiday Nursery. The nursery comprised seven acres and a home which they built in 1954. Their seven acres soon became a small private arboretum. Fred Bergman of Raraflora was a close friend of John's and gave him many dwarf plants for the property. John once began construction of a pit greenhouse, but changed his mind and filled the pit with soil. The resulting unsightly mound allowed Bergman to do his friend a favor. Bergman landscaped it with many choice plants but refused any sort of payment.

Temple University approached John in 1958 about the position of Chairman of the Department of Horticulture at the Ambler Campus. John accepted on a part-time basis and retained the position until 1968. He then assumed a position with the Barnes Foundation in Merion to teach propagation. After two years he resigned due to the deterioration of his speech. John was suffering from Parkinson's Disease, and its ill effects became more apparent as time passed.

In 1963, John formed a partnership with the Breadys, a landscaping and nursery business, and was made president of the new company. This relationship was dissolved in 1970. He, however, continued as a consultant to the company for several years thereafter.

John worked with hemlocks throughout his active life. One friend, Layne Ziegenfuss of Lehighton, Pennsylvania, was very important to him as the effects of Parkinson took its toll. Many hours were spent talking about plants and reviewing John's notes. John especially recalls sharing with the Ziegenfuss family many fine noonday meals, which were big enough to "feed an army". When John could no longer drive, Layne took him on several important plant excursions so he could continue his research.

Frank Mitsch of Maple Glen, Pennsylvania has been a very close friend of John's for over ten years. They traveled extensively on plant and other business. Their most memorable trip taken together with their wives was the International Plant Propagators Society meeting in the British Isles in 1975. The headmaster of a horticultural school was tour director and took the group to many gardens not open to the public as well as Kew Gardens and Hillier's Nursery. Despite the debilitation of Parkinson's which made walking difficult, everyone had to hustle to keep up with John. He even climbed onto the battlements of some of the castles they visited.

Encouraged by several good friends, John undertook the revision of his thesis on Canada Hemlock for publication. But his continued physical deterioration foreclosed the research necessary to complete the revision. A number of plantsmen then undertook the rewrite which John was unable to complete. All will have to agree that the completed text is a crowning achievement to a long, illustrious career.

As Layne Ziegenfuss has often told me, John is not a person to brag about his achievevents. I believe John felt they would speak for themselves, but he is very proud of many things accomplished. He made a collection of oak herbarium specimens for the Morris Arboretum. He created many beautiful landscapes in the Philadelphia area. One of his best jobs was done for his neurological surgeon, Dr. Richard Davis of Villanova, Pennsylvania. John is an accomplished photographer, which is evidenced in the pictures in his original thesis and text. His interest in and concern for his college students is reciprocated through the contacts he maintains with several.

One fact about John not known to most people is his color blindness. He once mistook a female holly for a male as he did not notice the red berries, a mistake his class found humorous. But most of the time John had an advantage in that he could detect many subtle shades of green overlooked by normally sighted plantsmen.

Among John's academic honors are membership in Kappa Phi Kappa (educational) and Society of Sigma XI (scientific). In 1967, the chairman of the Educational Committee for the International Shade Tree Conference asked John to exhibit photographs, with captions, of some outstanding trees at the conference in Philadelphia. This exhibit was well received and then put on display at Temple University, Ambler Campus. His awards of merit include the Buckley Award for the best educational exhibit at the Philadelphia Flower Show on behalf of Temple University in 1968. In August of 1970 at the annual convention of the International Shade Tree Conference John received an award of merit for "Science, Research and Preservation". During November, 1975, John was presented with an award of merit by the Pennsylvania Horticultural Society as "Outstanding Plantsman". The newly formed American Conifer Society has named John an honorary founding member for his outstanding contributions to the plant world.

John's interest in trees of historical importance led to the registry of twelve of the trees which he measured and photographed in southeastern Pennsylvania in the American Forestry Association's "Social Register of Big Trees". John's work in searching for big trees is reflected in his funding of such a project.

John Clymer Swartley is as big a man in spirit and actions among men as his hemlock is in a mature Pennsylvania-New England forest. I feel privileged to have met and talked with John, and I hope that as a result of reading his biography, others may also better know this dedicated plantsman.

Robert L. Fincham

April 24, 1983

JOHN C. SWARTLEY

1 INTRODUCTION

The Hemlocks (or more accurately, perhaps, the Hemlock Spruces) are an important genus of woody trees comprising about ten species and one natural hybrid. Their native habitat is confined to North America or to South and South-East Asia.

The four American species are widely distributed in their respective habitats and of these, one species, *Tsuga canadensis,* appears to be genetically very complicated. It is consequently very variable—phenomenally so—and this has given rise to a range of variants wider than perhaps can be found in any other tree species. Because these appear sporadically, they do not lend themselves to infra-specific botanical classification, but many of them make excellent garden plants, so it is surprising that hitherto no comprehensive study of this species has been published and that so little use has been made by landscape architects and home gardeners of the numerous garden forms—especially the slow-growing dwarfs—that are available. Since none of the other species exhibit this extreme variability the major part of this book has had to be devoted to the variants of *Tsuga canadensis.*

The native eastern or Canada hemlock, *T. canadensis* is the State tree of Pennsylvania and the Western hemlock, *T. heterophylla* is the State tree of Washington.

The purpose of the present book is to gather together as much information as possible and to give readers the benefit of recent study in up-to-date collections, both private and public. Many plants first recorded in 1939 have been identified, so it has been possible to study their behaviour over a period of forty years. No better evidence of their distinctiveness and reliability as garden plants could be hoped for. A large number of other variants have been examined, and of these a few are here named and described for the first time. Because the existing number of cultivars is large—possibly already too large—selection has had to be rigorous. Distinctiveness in any new selection is of course of prime importance: many of the existing cultivars are already so closely alike that identification is often very difficult. Other criteria have had to be: "Has it been propagated? Is the mother plant in a place from which it is likely to be properly distributed?"

Frequently a variant (seedling or mutation) is noticed, propagated and grown on by some keen-eyed rambler or gardener. In due course, when it has made an interesting little plant, a few cuttings are given to visiting friends (not always giving them the same name as on the label) or to a local nurseryman who raises and sells a few plants to selected customers. In this way a long list of untested, unregistered, unpublished and often unrecorded "names" has developed, often for plants lacking that vital claim to distinctiveness. Too much reliance on memory, faded labels and other hazards has resulted in the proliferation of misspellings and suspicious-looking synonyms. But since these unsatisfactory names are met with in collections and sometimes in nursery catalogues, many of them must appear provisionally in this book. In any revision, the reviser will have to re-examine every name, so the author would be glad to hear from anyone able to confirm, correct or amplify present knowledge of these waiters in the wing.

The author's aim is to make this book interesting and useful to horticulturists of all kinds, as well as to the specialist grower and the scientist.

ORIGIN OF THE NAME *TSUGA*

The name *Tsuga* first appeared when Stephan I. Endlicher (*Synopsis coniferarum.* 83. 1847) an Austrian botanist, applied it to a section in the genus *Pinus*. A few years later Elie-Abel Carrière (*Traité générale des Conifères.* 185. 1855) a botanist at the French National Botanic Gardens, appropriately adopted the name *Tsuga,* when he isolated hemlocks as a separate genus.

The common names of *Tsuga canadensis* are fully recorded by George H. Sudworth (*Check-list of Forest Trees,* 27. 1927), with the states in which they are used:

Hemlock: Maine, New Hampshire, Vermont, Massachusetts, Rhode Island, Connecticut, New York, New Jersey, Pennsylvania, Delaware, Wisconsin, Michigan, Minnesota, Ohio, Ontario (Canada).

Hemlock spruce: Vermont, Rhode Island, New York, Pennsylvania, New Jersey, West Virginia, North Carolina, South Carolina and England.

Spruce: Pennsylvania, West Virginia, North Carolina.

Spruce pine: Pennsylvania, Delaware, Virginia, North Carolina, Georgia.

Pine: Pennsylvania.

These names are additional to the names Canadian hemlock and New England hemlock in common use.

The first mention of the hemlock was made by two early French explorers; Jacques Cartier, who in 1534 used the name *d'if* meaning yew, and Samuel Champlain who used the word *pruce* in 1604, meaning spruce (Marie-Victorin, 1926).

Pierre Boucher (1664) furnished a brief description of the *pruce*. He stated that it is usually a large tree without branches for 30 or 40 feet, thick and rough bark, and wood which does not decay as readily as the wood of certain other trees.

The New York Indians used the descriptive name *Oh-neh-tah,* which translated means "Greens on the stick". Sudworth does not give the authority for *Oh-neh-tah,* but it sounds like *Hoaonaadia,* pronounced *Hoe-na-di-a* or *Hoe-o-na-di-a* according to Isaac Lyon, Head Chief of the Onondaga Indians, in an interview in 1939 with the author. It is the universal Indian name for hemlock. The north country (now Canada) was originally called *Hoanaadia,* literally land or place of hemlock. Canada hemlock is thus a combination of an Indian name and a middle English name applied by Europeans, because the graceful spray of the tree reminded the English colonists of the poison hemlock, *Conium maculatum,* a weed of the carrot family. An alternative suggestion sometimes put forward is that the leaves of some species when crushed have a scent like the poison hemlock of Europe. The philosopher Socrates was put to death by the "hemlock cup." No part of the hemlock tree is poisonous.

The origin of the name *Tsuga* was explained in *Hemlock Arboretum Bulletin* 8. 1934. There Mr. Jenkins wrote:

> To Miss Helen B. Chapin, a distant relative and a talented oriental scholar who has lived in China and Japan, I am indebted for much of the following information regarding the derivation and meaning of the Japanese word 'Tsuga'.
>
> It is a remarkable fact that the Japanese apparently had no written language of their own, writing having been introduced from China about A.D. 400. For a long time Chinese writing was confined to scholars, government officials and the nobility. The introduction of Buddhism in A.D. 552 hastened its more general adoption, but even in later epochs, it did not spread far outside court and ecclesiastical circles. Thus Chinese, having been the literary medium of Japan ever since its introduction, used in much the same way as Latin was used in Europe, has contributed to the formation of the Japanese language both spoken and written. This has been attained

despite the extreme dissimilarity of the two languages, which are at opposite ends of the pole so far as construction is concerned.

The Chinese characters which it is thought described the hemlock mean "tree 椤木 with hanging branches," which would be an accurate description. There are in Japan several hundred written characters which the Japanese have themselves originated and among them is the word 栂 which is composed of the elements "tree" and "mother" and which is pronounced "Tsuga." This sound probably existed as a Japanese word long before the introduction of Chinese writing. "Tree mother" or "mother tree" would also be an appropriate and sentimental attribute and description for the hemlocks of all countries.

It should be stated that Kamo no Mabuchi, a great Shinto scholar (1697-1769), thinks the Chinese word given above may have referred to the box tree and not the hemlock but we all know that "tree with hanging branches" would much more fittingly describe the latter than the former no matter where they grow. In early Japanese poems this Chinese character is used as a "pillow word" or pun because of its sounding like "tsugi" which means "following" or "continuing" and hence the word "tsuga" is used to wish the Emperor a long and happy rule.

We therefore conclude that "tsuga" is a word dating back many centuries in Japan, that it was originated there, that it is applied by the Japanese to their hemlocks, that it means "tree mother" and that it is now used as a Latin word describing the genus which includes the ten or more species of conifers known in English as hemlocks.

ECONOMIC USES

Two uses of hemlock are on record from early contact with the Indians. André Michaux (*Flora Boreali Americana* 206. 1803) records that certain tribes made tea from hemlock needles, thereby obtaining enough vitamin C to prevent scurvy. In the reference collections in the Department of Economic Botany at the Royal Botanic Gardens at Kew, England, there is a sample of bread made from the inner bark of the hemlock (probably *T. mertensiana*). The following account of how the Indians made it is told by Alexander Mackenzie, (the explorer for whom the great Mackenzie River is named) in his book *Voyages . . . through North America . . . in 1789 and 1793* (Vol.2. 211. 1802):

In July 1793, as his party approached the Pacific Coast they noticed numerous hemlock trees that had been stripped of their bark. They came to the village of a friendly chief who entertained them with a feast. Among the victuals was something that resembled the inner rind of the cocoanut but of a lighter colour. This he dipped in oil and having eaten it indicated by his gestures how palatable he thought it. He then presented me with a small piece which I chose to take in the dry state. A square cake of this was next produced, then a man took it to some water near the house and having thoroughly soaked it he returned and, after he had pulled it to pieces like oakum, put it into a well-made trough, almost three feet long, nine inches wide and five deep: he then plentifully sprinkled it with salmon oil, and manifested by his own example that we were to eat of it. I just tasted it, and found the oil perfectly sweet, without which the other ingredients would have been very insipid. The chief partook of it with avidity, after it had received an additional quantity of oil. This dish is considered by these people as a great delicacy; and on examination I discovered it to consist of the inner rind of the hemlock tree taken off early in summer and put into a frame, which shapes it into cakes of fifteen inches long, ten broad, and half an inch thick; and in this form I should suppose it may be preserved for a great length of time. This discovery satisfied me respecting the many hemlock trees which I had observed stripped of their bark.

Other early uses for the hemlock are referred to by John Josselyn, an English gentleman who wrote a book about his travels in New England in 1663. In his book *New England Rarities* (64. 1672) he states:

> Hemlock Tree a kind of Spruce: the bark of this tree serves to dye Tawny; the Fishers tan their Sails and Nets with it. . . . The Hemlock tree is a kind of spruce or pine, the bark boiled and stampt till it be very soft is excellent for to heal wounds and so is the Turpentine thereof.

BARK USED FOR TANNING

The tanning industry was largely responsible for decimating our stands of virgin hemlock. Hemlock bark is rich in tannin and the bark of the American species was formerly used very extensively in the tanning industry. Western hemlock is richer, containing 10-12 per cent, to only 8-10 per cent in the Eastern species. With the reckless use of natural resources that has characterized pioneers since man first began exploring and spreading around his habitat (criminal folly from which mankind is still not free) large areas of primeval forest were virtually destroyed for the bark, this being stripped, the trees being left, unused, to rot. Hemlock bark at one time supplied two thirds of the requirements of the tanneries of western North America.

In 1938 (*Hemlock Arboretum Bulletin 24*) Charles Jenkins wrote:

> Walking through the woods near our summer home in the Poconos the other day I came across a giant hemlock log, half decayed yet showing its great length and circumference. It had been cut years ago for its bark and the timber allowed to rot. This section in the latter part of the last century was a great tanning centre and the cutting of hemlock and chestnut bark an important industry. The little hamlet of Canadensis, four miles from the railroad, had a good sized tannery with which Jay Gould, the railroad financier, was somewhat connected. Tannersville, another village twelve miles away, received its name from its most important industry. As it takes several tons of hemlock bark to tan one ton of leather, it was cheaper to transport the hides to these mountain villages than to send the bark out to where the hides were produced. But the tanneries have long since gone. In the 1890's there was one remaining in Stroudsburg, our home town, and an interesting feature of it was the immense piles of bark built up regularly like a row of dwelling houses and waiting to be used. I have had in my file for several years a thoughtful and comprehensive account of the use of hemlock for tanning, written by Mr. Fraser M. Moffatt, President of the Tanners council of America. Coming from a long line of tanning ancestors he knows the industry from the ground up. This is what he says:
>
> > The scientific development in the tanning industry during the past half century has gradually pushed hemlock bark to a relatively minor position from what it occupied in prior years. The bark of the hemlock tree, varying in its amount of tanning content according to geographical location between the ranges of eight and 12 per cent, is no longer the important factor which it was in the latter part of the last century. It has been superseded to a large degree in what we know as the Heavy Leather field (sole leather, etc.) by extracts of oak wood and many other vegetable components. The bark itself, however, where procurable at reasonable prices, relative to other prices, ranging in figures from $8 to $20 per ton, is held desirable and useful but not entirely essential. Where hemlock forests were prolific, particularly in the eastern part of the United States, the cheapness of the material used created a very large business in the leather made therefrom. Much of this leather was for export. Later years, with the destruction of these forests, saw the employment of other vegetable tannins and later the use of mineral tannins. The result has been that the leather today is better adapted to the needs of the population and probably of higher grade than anything which this country has ever seen. There is a definite shrinkage therefore, due to the unavailability of supplies, of tree barks for tanning purposes.

Hemlock bark comes mainly at present from Pennsylvania, Wisconsin, some parts of West Virginia and North Carolina. A residue is still left in Michigan. There have always been reserve supplies on the Pacific coast of enormous quantities of western hemlock. The difficulty of transportation to the tanning centres which are mainly east of the Mississippi River, is now being met by an endeavor to make a tannin extract from the Pacific coast hemlock, *T. heterophylla.*

The bark of the hemlock tree, in common with the bark of chestnut oak up to the later part of the 19th century, was important as a basis for the tanning industry in heavy leather. Tanners and lumbermen were a menace to the hemlock forests. That situation no longer exists, due to scientific progress. The hemlock trees are no longer cut simply for their bark. Hemlock, like most other timber of its type, has now its main market for pulp wood, not for manufactured timber. Bark is incidental and where possible is saved and sold.

Dr. David Bailey of the Northeast Region Research Center, U.S.D.A. A.R.S., Hides and Leather Laboratory, Philadelphia, Pennsylvania, has kindly brought this information up to date. He reports that industrial tanning in 1982 does not utilize Hemlock. The industry does not do any "vegetable leaching" any more—all tanning is done by other chemical and mechanical processes. No leather manufacturers or processors in the U.S. in 1982 use hemlock, either on the east coast or the west coast.

He added that circa 1820-1860 all leather tanning in the Adirondacks in New York State (the center of the tanning industry at that period) used *Tsuga canadensis* exclusively. He pointed out also that one major tanning company in the late 1920's on the west coast (ie. northern California, Oregon and Washington) used *Tsuga heterophylla* exclusively. But this was the last company to use it, and when the company went out of business also in the late 1920's—commercial use of *Tsuga heterophylla* for tanning ceased.

He did, however, add that a few private individuals, possibly in the very rural areas of New England, might still use the bark of *T. canadensis* as sort of folk-art. Likewise on the west coast *T. heterophylla* (with the "back to nature" movement and the resurrection of Folk Arts and Crafts) is probably still used for tanning—but on a very limited scale.

HEMLOCKS FOR LUMBER

Of our eastern species, neither *Tsuga caroliniana* nor *Tsuga canadensis* plays a major role in the lumber industry, the former because its availability is limited by the restricted habitat of the species, the latter because it is inferior to the western species in almost every respect. It produces timber which is buff to light brown in colour; without distinctive taste or odor (when seasoned); weighing from twenty-five to thirty-five pounds per cubic foot, difficult to season and poorly resistant to decay; with a pleasant, close grain with the growth rings distinct. But in carpenters' language it "works badly", being unevenly and often cross-grained, dry, brittle and harsh under hand or machine tools and subject to ring-shake caused by the wind which means that many boards are cracked and unusable.

In the early days it was used locally for building purposes and many of the old barns were constructed of hemlock siding. Lumber production in these species reached its peak during the closing years of the nineteenth century and the opening years of the present one. It still finds local use for such purposes as packing cases, railway sleepers, structural uses and carcassing, but its principal use is, and will probably remain, for pulping for the paper industry. Although, there is some evidence that the demand for lumber is on the increase, currently the eastern species are valued highly as cover and winter shelter for white-tailed deer and other fauna and for this reason cutting is restricted in some areas.

The western species of hemlock, *Tsuga heterophylla,* on the other hand, is now a major source of lumber, although it seems historically to have been slow to attain its present position, perhaps from its association with the defects Eastern species suffer.

Sir John Maxwell in the Report of the Conifer Conference held by the Royal Horticultural Society in London in 1932 recommended this species for planting in Britain. He wrote:

> The timber of the western hemlock, *T. heterophylla,* was for long neglected in its own country, as well as here, but has now come into its own and is in great demand for building and all kinds of indoor work. Its quiet colour and nice texture are extremely attractive. These merits, in a well organized market would make up for its slow growth. It is extremely hardy and has hitherto been quite free from disease. Like other fibrous-rooted trees it grows well in peat. Altogether it seems destined to play a more important part in our woods than that hitherto assigned it.

Sir John was of course writing of a need at that date for a "well organised market" to be developed in Britain. In this country there is certainly no such lack to-day, but the growth of the industry in the past has been slow. Charles Jenkins, in *Hemlock Arboretum Bulletin* 42. 1943, quotes an interesting letter from E. E. Carter, then Chief of the Division of Timber Management of the Forest Service of the U.S. Department of Agriculture:

> Possibly you may be interested in the increased use of western hemlock, *T. heterophylla,* brought about by the urgent needs for wood in war activities. Western hemlock has been a source of high grade wood pulp for years, and has found some considerable use as lumber. The war demand for structural lumber greatly increased that use, and studies and experiments by the Forest Products Laboratory at Madison, Wisconsin, led to modification of the specifications of lumber for use in building airplanes—trainers, gliders and the like rather than combat planes—to admit western hemlock of the requisite clearness, straightness of grain, density, etc. The W.P.B. specifications with respect to hemlock of airplane grade, is that the tree must have grown on land above 1500 feet elevation and it must have at least six rings to the inch of radius, preferably more than six. This is really the equivalent of telling producers not to expect to find trees having wood with the requisite number of rings per inch of radius and the requisite density except above that elevation in western Washington. The tree grows too fast on the good sites of the lowlands.

The timber from the Western hemlock, *Tsuga heterophylla,* is light yellowish-brown, often with a tinge of red; without taste or odor (when seasoned), weight between twenty-five and forty pounds to the cubic foot; easy to kiln dry but not durable in unfavourable situations. It is harder than timber from the Canadian hemlock, is comparatively free from ring-shake and it machines satisfactorily. It is therefore not surprising that it has become one of the major timber trees of the world.

Robert L. Etherington, Director at the Pacific Northwest Forest and Range Experiment Station of the U.S. Department of Agriculture at Portland, Oregon, has kindly supplied some interesting statistical information about the production of Western hemlock in the several western States and in British Columbia in Canada.

In 1976 (the last year for which full figures are available) the total consumption recorded for the different outlets was as follows:

As lumber	1,554,758
Veneer and plywood	274,508
Pulp and Board	410,022
Foreign log export	1,484,092
Post, pole, piling	695
	3,724,075*

*The figures are given in "Thousand board feet, Scribner Scale." Except to those in the industry the figures will mean very little, except as a means of comparison.

As can be seen, lumber is the primary domestic manufactured product. Foreign log exports go chiefly to Japan and Korea. Production and the market price varies considerably according to world conditions, being at present in recession, but there appears to be an increase in demand because in 1982 mainland China was purchasing large quantities of this species in log form. The Japanese convert the logs into lumber for house and general construction work. They value the aesthetic value of the wood left untreated.

The mountain hemlock, *Tsuga mertensiana* is of little importance as a forest tree. Where it is cut the log is marketed with the Western hemlocks. In general, being a mountain species, the stands are inaccessible and very often of poor quality—very knotty. So its use is limited to pulping.

Mr. R. M. Fredsall, Resources Director, Western Wood Products Association, also of Portland, Oregon, has also kindly supplied annual figures of production for the years 1970 to 1978, which confirm that the general trend in demand is steadily increasing. In view of this fact and the inability of many countries of the world to produce any timber whatsoever it is a relief to be able to record that the U.S. government (amongst others) is aware of the situation, so that the future is not being left entirely to look after itself. A Report *The Outlook for Timber in the United States* published in 1974 analysed the resources and the likely demands for the next few decades. It points out that "Growing concern over prospective depletion of non-renewable mineral resources, and the higher energy requirements and pollution impacts resulting from non-timber resources in lieu of timber products all emphasize the growing importance of timber in the U.S. economy. UNLIKE MOST COMPETITIVE PRODUCTS, WOOD IS A RENEWABLE INDUSTRIAL RAW MATERIAL".

FOREST TREE MANAGEMENT AND GENETIC STUDIES

The emphasis given to the last preceding sentence is mine, and although the subject of this section is outside the scope of the present work attention may here be appropriately drawn to the work that is being done, both in this country and in Canada on the scientific and professional management of commercial forests to increase production of our main timber trees, including the Western hemlock. In this, both governmental departments and private agencies are involved. For any reader interested, the U.S. Department of Agriculture manual *Management of Western Hemlock-Sitka Spruce Forests for Timber Production* by Robert H. Ruth and A. S. Harris (General Technical Report PNW-88 June 1979) gives a full account of this activity from the practical standpoint, with sections on site preparation, regeneration from seed, programmes for thinning, fertilisation and protection of the crop and other relevant subjects. And in a paper *Tree Improvement in Western Hemlock*, Richard F. Piesch of the Canadian Forestry Service, Victoria, B.C., gives an interesting account of the experimental work that is being done on the genetic improvement of the species. Originally started in the field as simple selection of "plus trees" for seed collection, the work has been extended to laboratory studies of genetic variabilities within the species and the scientific application, through controlled pollination, of this knowledge to the development of superior strains and their widespread use in afforestation programmes. The uses and limitations of vegetative propagation are discussed and reference is made to recent successful production of plantlets by tissue culture. Since each of these publication has a good bibliography, they form a good introduction to a study of the subjects they cover.

OTHER HEMLOCK USES

USE AS PULPWOOD FOR PAPER-MAKING

By far the greatest use that has been made of the national heritage in the primeval hemlock forests has been, and still is, the pulping of the wood for paper-making.

Conservationists are very properly becoming increasingly concerned about "endangered species" amongst both plants and animals. What mankind should also be worried about is that within a very few generations *Homo sapiens* will be on the list. The difference in his case is that whilst the other threatened species have suffered from the loss of habitat or a suitable food supply, or other privations not within their control, this species faces extinction because of destruction of its environment by its own wasteful misuse of natural resources.

The importance, in this context, of such a simple commodity as paper was brought home to the author recently when browsing through a book *The Outlook for Timber in the United States* published by U.S. Department of Agriculture (Forest Resource Report No 20. 1974) where, on page 188 he read:

> Since 1920 pulpwood consumption in United States mills has increased 12 times . . . Export demand . . . has increased nineteenfold. As a result of such growth, nearly half of the cubic volume of timber harvested from domestic forests is used as pulpwood.

USE FOR FUEL

Hemlock wood has always been widely used for burning as fuel, a use for which it is very suitable, provided it has been felled sufficiently long for the sap to have dried out. Used fresh it is a source of irritation and danger of causing fire from the minor explosions due to the sudden conversion of the wet sap into steam.

Cross section of a Canada hemlock trunk, showing the effects of repression due to early competition from surrounding trees. Notice the cracks, which often occur in second growth due to windshake. Photo by William M. Harlow.

HEMLOCKS IN VIRGIN FOREST

Hemlocks are very long-lived and, although slow growing, are capable of becoming very large trees. Annual growth ring counts on stumps have revealed an age of eight or nine hundred years.

In the east at the present time, Canada hemlock lumber is no longer of importance. Nearly all of the virgin stands have been cut down and regeneration has generally been unsatisfactory. Hemlock is a climax tree, so after-cutting conditions are no longer favourable for seedling production, and forestry planting of nursery-grown seedlings has not been really successful. This, coupled with slow growth, has limited hemlock production in the forest. A few virgin stands of large specimens of eastern hemlock may be observed in areas of primeval forest remaining in various parts of the country. Charles Jenkins in *Hemlock Arboretum Bulletin* 52 (1945) printed the following extract from "Pennsylvania Park News" published by the Pennsylvania Park Association, Philadelphia, Pennsylvania, listing some of the parks established in that State where isolated specimens or stands of towering trees are to be found:

> The Joyce Kilmer Forest Monument, Union County, comprises about twenty-one acres of old hemlock, some white pines and many magnificent hardwoods situated on a slope of White Mountain in the Bald Eagle Forest.
>
> Cook Forest Park has several thousand acres of white pine, hemlock and hardwoods along the Clarion River at the junction of Forest, Clearfield and Carbon counties.
>
> Ricketts Glen, thirty miles west of Wilkes-Barre, located in the gorge of Kitchen Creek below Ganoga Lake, in Sullivan County, is lined with grand original hemlocks and hardwoods for a distance of five miles and in the gorge is a series of twenty-nine beautiful waterfalls.
>
> The Alan Seegar State Forest Monument, located along Stone Creek in Huntingdon County, has hemlock trees of unusual size and beauty, undoubtedly many of them standing in the same spot when Columbus first came to America. Within this area is a magnificent growth of rhododendron, some with forty-foot stems.
>
> Martin's Hill State Forest Monument has nine acres of old original hemlocks, within a gorge in the highest mountains of Pennsylvania. This monument is located at the foot of Buchanan State Forest. It is 3075 feet above sea level.
>
> Synder-Middlesworth State Forest Park, located in Snyder County about five miles west of Troxelville, has approximately 250 acres of original hemlock, pine and hardwoods.
>
> Detweiler Run State Forest Monument, comprising about fifty acres of giant original forest trees, mostly pine and hemlock, has an almost impenetrable understory of rhododendrons. Students of early forest conditions declare that it is possibly the truest picture of a primeval forest in the state. It is located near the headwaters of Detweiler Run, Huntingdon County, in the Logan State Forest.

In a letter of March 2, 1976, William S. Corlett of the Department of Environmental Resources of the State of Pennsylvania, stated that the hemlocks are alive and well in all of the Parks just mentioned. Mr. Corlett also added that virgin hemlocks can also be found at the following locations: The Hemlocks, Perry County; Heart's Content on Allegheny National Forest; Salt Springs, Susquehanna County; Tionesta Scenic Area, Forest County; McConnell's Narrows, Union County; Mount Logan, Clinton County; Bark Cabin Run, Lycoming County; Picture Rocks, Lycoming County.

In other States throughout the natural range of the species, i.e. New Brunswick, Nova Scotia and Quebec in Canada, and in this country south to Maryland and west to Eastern Minnesota, with disjunct populations occurring in Indiana and Ohio, and south along the Appalachian Mountains to Georgia and Alabama, there do occur preserved tracts of old climax trees. Such relict populations are (in some States, including

Connecticut, Maine and Massachusetts) frequently in areas preserved by municipal, county or even State authorities. Dr. Dudley tells me that in northern New England, particularly around freshwater lakes in Hancock County, Maine, he has noticed many trees 2-4 feet diameter (at breast height) all situated on private property.

In the *Bulletin of the American Association of Botanical Gardens and Arboreta* (**16** (1):47. 1982) Michael Ruggerio, Assistant Foreman of Garden at the New York Botanical Garden gives an interesting account as follows:

> At the heart of the Garden lies a 40-acre area known as the Hemlock Forest, the last uncut woodland within the boundaries of New York City. It has been described as "the most precious natural possession of New York City."
>
> The Hemlock Forest exists as a post-glacial relic of the extensive coniferous forest which covered the Atlantic seaboard following the retreat of the last ice age, some 20,000 years ago. As such, it represents the southernmost surviving representative of this community.
>
> The existence of the Hemlock Forest is in part due to the cool, moist microclimate associated with the steep west bank of the Bronx River. This southernmost occurrence may also be related to the outcrop of gneiss and other ancient crystalline rock-types at the Garden which, further south, are buried beneath a sandy glacial outwash plain. Its persistence, however, is largely due to the ownership for more than one hundred years by the Lorillard family as part of their home and tobacco manufacturing site. It simply was not cut at a time when most of the rest of the area was cleared for farm and home sites. In the late nineteenth century, the land was deeded to New York City for park purposes, and, in 1895, was designated as part of the New York Botanical Garden.
>
> The Hemlock Forest has been in a state of decline for at least the past 30 years. Uncontrolled public access has led to soil compaction, vandalism and fires; all of which encourage faster and more vigorous hardwood and alien trees, such as *Phellodendron amurense,* to encroach. Eradication of herbaceous plant species which help build and maintain the soil and that create the moist, protected conditions for successful germination of hemlock seeds is another problem. Also, a high local population of rodents eats hemlock seed cones before they are ripe, and pedestrian traffic tramples the few seedlings that do germinate.
>
> As steward of the forest, The New York Botanical Garden has commenced a program to restore and rehabilitate the stand.

HEMLOCK CHAMPIONS

Some outstanding specimens have been recorded in the past. Many of these were collected by Mr. Jenkins, and for those interested, a browse through the issues of the *Hemlock Arboretum Bulletins* would be time well spent.

A *National Register of Big Trees* is maintained by the American Forestry Association of 1319 Eighteenth Street, N.W., Washington D.C. 20036, who have kindly permitted me to extract the following details from their records. There seems to be as much competition amongst big trees to be recognised as a champion as there is amongst athletes and all the other cranks who set out to establish records of one kind or another, so these details are included here for reader interest only.

The "vital statistics" of each tree are recorded as three figures, these being the trunk diameter at 4½ feet above ground, the height in feet and the span of the tree in feet. Where the date of measurement is known the date is given in brackets.

Tsuga heterophylla (Western Hemlock)

A tree in the Olympia National Park, Washington, reigned as National Champion from 1957 to 1978. In 1954 its size was recorded as being 27′ 2″/125′/47-6. In 1972 fresh measurements record it as 27′ 4″/164′/62′. It was evidently still growing well but in

1980 it had to concede co-championship to a tree on private land about 4 miles south of Tillamook, Oregon, and the title is still being jointly held. The Tillamook tree is on record as measuring (1978) 25′ 5″/195′/50. Unsuccessful claimants are recorded from Oregon and Idaho.

Tsuga mertensiana (Mountain Hemlock)

In this species the laurel has been consistently won ever since 1956 by a tree in California, but claims have also been made on behalf of big trees in Oregon, Washington and Idaho.

		Date	Girth	Height	Spread
1956-1982	Tree in Stanislaus National Forest, Alpine Co., California	1955	21′9″	118′	83′7″
		1972	23′1″	163′	—

Tsuga caroliniana (Carolina Hemlock)

As must be expected of this species, which has a very limited natural habitat, all claimants to recognition as Champion come from a single State. The records so far are:

		Date	Girth	Height	Spread
1956/61	Tree at Dunn's Rock, North Carolina		9′4″	—	—
1966	Tree near Brevard, North Carolina	1952	9′4″	—	—
1967/82	Tree at Linville Falls, North Carolina	1966	9′9″	101′	54′

Finally, in *Tsuga canadensis* (Canada Hemlock) as befits a species so widely distributed in the wild, claims have been submitted on behalf of trees in the States of Maryland, Massachusetts, Michigan, New Jersey, Ohio, North Carolina, Pennsylvania, Tennessee and West Virginia, but all to little avail since a tree in Tennessee was acclaimed National Champion when the Scheme first began and it reigned unchallenged until 1980, since when it has had to share the honor with another tree recorded in West Virginia.

		Date	Girth	Height	Spread
1945-1978	Tree in Great Smoky Mountains National Park, Tennessee	1945	17′9″	—	—
		1949	19′9″	98′	69′
		1972	19′11″	100′	70′
1980-1982	Co-champion with the above, a tree at Aurora, West Virginia	1979	18′8″	123′	68′

HEMLOCKS IN THE BRITISH ISLES

The first mention of hemlock in England was by Plukenet (1692, p. 21) who referred to a tree in the garden of Bishop Henry Compton. This had probably been sent over by John Bannister, a Virginia naturalist. Another introduction was described in detail in *Hemlock Arboretum Bulletin* 2. 1933.

> Sometime before 1730, Dr. Christopher Witt, a Germantown physician and plantsman, sent his friend and correspondent Peter Collinson, the famous English botanist, some hemlocks which were planted in his garden at Peckham in Surrey. In 1749, Collinson's collection was moved to Mill Hill at Hendon. Still growing at Mill Hill, now a boarding school for boys, is a Witt hemlock now well over two hundred years old. It was measured by Loudon in 1835, each of its two trunks was a foot in

diameter and about fifty feet in height. In a recent letter, Major N. G. Brett-James, author of the life of Collinson, wrote about 1932, "The Hemlock Spruce is still in good health and is not changed in size from when Dr. H. Harris measured it in my company for Dr. Hingston Fox (1923) and so it is almost exactly the same as in Loudon's time (1835).

The explanation for this failure to increase in size may be found in *Trees of Great Britain and Ireland* where (Vol. 2. 241. 1907) of this tree Professor Henry wrote that it "still survives, but was, when I saw it in 1906, in poor condition, the soil being too dry for it".

Not surprisingly, the tree has long been dead.

The development of interest in the hemlocks in the British Isles can be readily followed because the Royal Horticultural Society holds a Conifer Conference every forty years or so. The first was held in 1891 (the Report constitutes Vol. 14 of the Society's Journal for 1892). In it were given the dates of introduction as follows: *T. brunnonia* (now *T. dumosa*), 1830; *T. canadensis,* by Peter Collinson in 1736; *T. mertensiana,* by Jeffrey in 1851. Other species mentioned but not dated are *T. caroliniana, T. diversifolia, T. pattoniana* and *T. roezlii* (both now included in *T. mertensiana.*) The late George Nicholson, then Curator at the Royal Botanic Gardens at Kew and Editor of *The Illustrated Dictionary of Gardening* wrote, "I have seen *T. canadensis* do well near water; in a wild state it is found in rather dry, rocky situations and generally on the north side of hills." No reference to the genus appears in the tabulated Statistics of Trees collected from the main collections and arboreta in Britain, because it was not one of the twenty-four species listed in the Questionnaire (on the returns from which the Statistics were based) as being of special interest.

The Second Conifer Conference was held in 1931, and on this occasion the Report was separately published with the title *Conifers in Cultivation* (Editor: F. J. Chittenden) 1932. The remarks by Sir John Stirling-Maxwell on the value of *T. heterophylla* as a timber tree in Britain have already been quoted (See p. 14). In a paper by Murray Hornibrook on dwarf conifers the *Tsuga* find no mention, but A. D. Slavin, then Horticulturist in the Bureau of Parks, Rochester, New York, in a paper on conifers cultivated in the United States describes four cultivars of *T. canadensis* and mentions three other species—*T. caroliniana, T. diversifolia* and *T. sieboldii.* The results of a questionnaire are set out in statistical form, and this time records of *Tsuga* trees appear in about eighty of the returns. A list is also given of the location and size attained at that date by the best trees in Britain of each species.

The Third and most recent Conifer Conference was held in 1970. Again the Report was published separately with the title *Conifers in the British Isles* (1972). As on previous occasions it contained up-dated statistical information collected from noteworthy collections of conifers, but as the emphasis this time was more on the largest specimens and their measurements, so far as the hemlocks are concerned the entries are mainly confined to *T. heterophylla,* the only species that has taken kindly to life in Britain. It is interesting to note that several references were made in the Report to the fact that this species is now regenerating freely from self-sown seed.

The conifer statistics were contributed by Alan C. Mitchell of the British Forestry Commission. In the same year the British Government publishing organisation, Her Majesty's Stationery Office (H.M.S.O.) published *Forestry Commission Bulletin No. 33,* also by Alan Mitchell, and also (most unfortunately) bearing the same title *Conifers in the British Isles.* This publication (*Bulletin* seems hardly the right term to use for a book that runs to 322 pages) is a mine of information on conifers that make it a work of general reference of value far beyond the British Isles. With the consent of H.M.S.O., Mitchell's *Key to the Genus* and adaptations of his descriptions of each species, together with the excellent line drawings by Christine Darter are reproduced in this book (See Chapters 3 and 6). Appended to each description in the *Bulletin* are lists of outstanding specimens in the British Isles, with location, date of planting (where known) and recent measurements. These lists bring the record of hemlocks in the British Isles right down to the present: they well repay study by anyone interested in the subject.

2 *HEMLOCKS OF LAWN AND FOREST*

> *The pine is called the kingly tree,*
> *It well deserves full royalty.*
> *The hemlock's called the princely tree,*
> *This lower rank is not for me.*
> *Right here and now O! Hemlock tree*
> *A royal crown is given thee.*
> *Not prince, but equal, queenly tree,*
> *In beauty, grace and symmetry.*
> *Great honors his, full honors hers,*
> *The King and Queen of conifers.*
>
> Charles F. Jenkins

There can be no doubt that a mature hemlock tree in a good, open spot is one of the most beautiful, perhaps the most beautiful sight in the world. Professor Charles S. Sargent, first and long-time Director of the Arnold Arboretum, Jamaica Plain, Massachusetts, in 1923 placed Carolina hemlock first in his selection of the "seven best conifers." (See *Heml. Arb. Bull.* 42. 1943). Woodbridge Metcalf, one-time Extension Forester of the University of California at Berkeley regarded the Mountain hemlock as the most outstanding. (See *Heml. Arb. Bull.* 43. 1943). And in the issue of "Plants and Gardens" devoted to conifers (*Brooklyn Bot. Gar. Record* **25** (3): 4. 1969), Professor Henry Teuscher's article entitled *The Fifty Finest Conifers,* included three hemlocks: *T. caroliniana,* Carolina Hemlock; *T. diversifolia,* Japanese hemlock and *T. heterophylla,* Western hemlock.

In the same handbook, other horticulturalists of note gave the ten best conifers for each region. *T. canadensis* is included for eastern Massachusetts, Long Island, New York City, Philadelphia, Washington D.C., northern Ohio, southern Michigan, northern Illinois and southern Ontario. Carolina hemlock is recommended only for one region, Long Island. In the West, Western hemlock, *T. heterophylla* is recommended only for northern California and the Vancouver area of British Columbia. Mountain hemlock, *T. mertensiana,* is listed only for the Puget Sound area of Washington State, although it undoubtedly grows well in other areas.

This attests to the fact that, although Carolina hemlock may be the most beautiful, Canada hemlock is the more popular and adaptable. The latter species with its wide distribution had a good head-start over the former, and also over the two Western species, *T. heterophylla* and *T. mertensiana.* These last two species were for long much confused in the nursery trade and even now, in the West, they are much less popular for landscape work than *T. canadensis* is in the East.

THE VALUE OF HEMLOCKS FOR LANDSCAPING

We can go back to one of our earliest Presidents for evidence of the use of Eastern hemlock as a landscape plant. In his Garden book, Thomas Jefferson on April 6, 1804 wrote, "Planted 40-odd hemlocks and Weymouth pines near the aspen thicket".

The hemlock was early recognized to be admirably adapted to landscape use. This is suggested in a quotation from *"Rural Essays"* written by A. J. Downing in 1854:

> We place the hemlock first, as we consider it beyond all question the most graceful tree grown in this country. There are few of our readers who have the least idea of its striking beauty when grown alone in a smooth lawn: its branches extending freely on all sides, and sweeping the ground, its loose spray and full feathery foliage floating freely in the air, and its proportions full of the first symmetry and harmony.

Years later Professor Sargent, one of our greatest plantsmen, confirmed Downing's opinion when in his *"Silva of North America"* (**12**: 61. 1898), he wrote "No other conifer surpasses the hemlock in grace and beauty".

In an article entitled 'The Eastern Hemlock and its Varieties' published in *Arborists' News*, **3** (4): April 1938, the author wrote as under. His remarks remain as true today as when they were written:

> Our native hemlock has long been loved by the common people. With its sweeping branches and its delicate spray it suggests friendliness, and the soft and feathery foliage invites us. This is in contrast to the stately pine which we admire and respect, but usually from afar. Mr. Henry W. Shoemaker of Clinton County, Pennsylvania, confirms this, for as a veteran collector of folklore, he states that he has found more legends concerning hemlocks than pines, probably because the former are closer to the common people.

Although the Canadian hemlock has not proved successful for re-afforestation its use in landscaping schemes and the home garden (along with the other species of hemlock) has few equals and no evergreen tree to surpass it. The plant purchased by the home owner or landscape consultant will have been raised from seed and will already have spent its difficult infancy in the nursery, so after a year or two to settle down it will begin to grow rapidly.

Such a variable species will, not surprisingly, vary from plant to plant in the seed-beds and so care in selecting plants with particularly attractive habit or foliage can be rewarding. Information on culture will be found in Chapter Eight: all that need be said here is that hemlocks prefer a deep, cool loam (What plant doesn't?) but that they will succeed in anything reasonably described as "garden soil", and also that, whilst hemlocks thrive in full sun, they are remarkably tolerant of shade.

The usual mistake made is in not realising that the pretty little shrub you have just turned out of its 5 gallon container hopes to live for several hundred years and become a tree 30 m or more high, with an equal or possibly greater spread. Of course you, the planter, will not live to see this development, but a hemlock, if it takes to you, will increase in height (and, correspondingly, in spread) by well over one foot annually. So please give it space! Especially is this true of the 'Sargent' hemlocks, whose spread will be much more than their height. They look such attractive "garden plants" when bought, but planting such a tree unless you can guarantee it fifty feet clear space in every direction in perpetuity is asking for trouble, anxiety and—eventually—anguish for the owner. Not you perhaps, or the chap you sell to, but someone, eventually.

In the modern landscape, the hemlock undoubtedly has greater use in mixed or mass plantings than as a specimen. There seems to be a steady trend away from the "specimen" idea. But here, too, we have a plant admirably suited for such use. The hemlock is graceful, it grows well in shade, it does not grow out of bounds quickly and can be restrained for years by judicious treatment; its dark green color is effective in mass and blends well with other plants. In fact, there are only two drawbacks—it dislikes city smoke and it does not tolerate extreme heat and dry soil such as one finds in the Prairie States. But fortunately here in the East, the city atmosphere is practically the only restricting factor. Of course, let it be thoroughly understood that the hemlock thrives best in a moderately moist situation and in partial shade, although it is satisfactory in other locations. This is merely a lesson from its natural habitat.

The variability of this species has enabled selections to be brought into cultivation having some particular claim to distinction, but for ornamental planting where there is room for the development of a large specimen, unselected seedlings can be expected to be as successful as any of the cultivated forms and these will have the better root systems and vigour associated with plants genetically raised.

When, however, the particular landscape value of one of the selected variations is to be made use of, vegetatively propagated clonal stock is usually resorted to, since although such variants are often regarded as "coming true from seed" there is usually considerable variation, often not becoming apparent until after many years of growth, by which time it will be too late if an unsuitable or unattractive tree has developed.

EFFECTIVE COMBINATIONS

The dark green hue and feathery grace of hemlock makes it an admirable background for flowering trees and shrubs. Whoever has seen magnolia, flowering cherries or dogwood against such a background, retains in his mind an indelible picture of the striking effect. Also, one can make wonderful combinations with rhododendrons and azaleas. Here again, nature sets the example. Let us quote a well known rhododendron expert, Mr. Joseph B. Gable of Stewartstown, Pennsylvania, "Have you noticed that *Rhododendron* and the *Tsuga* species are mostly found growing together in nature? That is, in general, wherever *Tsuga* is found native, the same parts of the world have their *Rhododendron* species and where one is absent, so is the other. To prove the rule, I can think of one exception: there are no *Tsuga* species native to the Caucasus, where several *Rhododendrons* are found. Conditions favorable to one genus seem to be also to the liking of the other."

WEEPING HEMLOCKS

Principal amongst the extraordinary variations typical of *T. canadensis* are the weeping or spreading forms usually lumped together in this country as "The Sargent Hemlocks". A full account of the characteristics, history and nomenclature of these trees, all consisting of four (or possibly five) clones originating from trees discovered in the wild and brought into cultivation towards the end of the last century will be found in Chapter Five (See p. 63).

Soon after the discovery of the Sargent hemlocks, propagations of one or more of the four original trees produced by grafting were being distributed by the former nursery firm of G. B. Parsons and Sons of Flushing, Long Island, New York. It was first listed in their catalog for 1879 where it was described thus:

> *Abies canadensis pendula Sargentii* Sargent's weeping hemlock, the most graceful and delicately beautiful evergreen known. When the leader is trained to a stake it can be carried to any reasonable height, each tier of branches drooping gracefully to the ground, like an evergreen fountain. It was first sent out from Flushing, having been received from H. W. Sargent, of Fishkill-on-Hudson.

There seems to have been a good demand for these trees and numerous survivors exist. The first public showing was at the great centennial exhibition in Philadelphia in 1876. Josiah Hoopes (*Book of Evergreens* 418. 1868) mentions the original tree at Wodenethe, home of the Henry Winthrop Sargent after whom these trees were later named. F. J. Scott (*The Art of Beautifying Home Grounds,* 549. 1870) writes of the weeping spruce as "an interesting addition to our stock of gardenesque evergreens. Will probably be a broad, flat-headed tree or large bush." In an English periodical (*The Garden* **8**: 310. 1875) an editorial introduces "The Weeping Hemlock Spruce. Mr. Samuel Parsons writes to us from Flushing, Long Island, praising the beauty of the Weeping Hemlock (*Abies canadensis* var. *pendula*) The ordinary form of the Hemlock is a very graceful tree, and this one will prove a welcome addition to our pleasure grounds." Later the same periodical (*The Garden,* **12**: 363. 1877) makes further mention of these trees and gives what appears to be the first illustration, a woodcut engraving made from a photograph sent in by Samuel B. Parsons of a plant then growing on his nursery. Also in 1875, in an appendix to the ninth edition of A. J. Downing's *'A Treatise on the Theory and Practice of Landscape Gardening'* Henry Winthrop Sargent refers to this "very interesting and distinct variety of Hemlock" and also mentions its availability only from the single nursery.

Somewhat remarkably, save for Hoope's reference to seeing one of the original trees, all these references are to the propagations coming from the Parsons nursery, and it is not until 1897 that Professor C. S. Sargent (*Garden and Forest* **10**: 449. 1897) published the first account I have been able to trace of their original discovery on the Fishkill Mountains. It has since then been repeated by many authors, but is now being challenged. (See p. 135)

A few of the survivors of those early propagations are listed here, State by State, but many others exist. Elsewhere (p. 136) I make the suggestion that a scientific attempt should be made to find a means for identifying the original clones, each of which I am now suggesting should be treated as a separate cultivar.

An example of staking, in the nursery of George L. Ehrle, Clifton, New Jersey, in 1938.

LARGE SPECIMENS OF SARGENT HEMLOCKS

MASSACHUSETTS

Brookline

At Holm Lea, the former home of Professor Charles S. Sargent.

In 1938 this tree was 6 feet high and 24 feet in spread. It forms a low mound and is branched to the ground. In 1965 it was 7 feet high and nearly 30 feet in diameter. This tree is No. 4 of the four original plants collected by General Howland and is the clonotype of the cultivar 'Brookline'. It forms a much lower and more wide-spreading tree than do the other clones. See p. 136.

T. *canadensis* 'Brookline', at Holm Lea, the former home of Charles Sprague Sargent, Brookline, Massachusetts.

Jamaica Plain

In the Arnold Arboretum of Harvard University.

A plant from an early graft (traditionally taken from the Matteawan tree) was measured in 1938, when it was 10 feet high and 30 feet wide. It has recently been badly damaged by vandals, having lost several main limbs. Since Tree No. 1 is available, Welch (Dwarf Conif. 323. 1965) was out of order in denoting this tree as the clonotype of the cultivar 'Sargentii', even if its clonal status can be verified.

Wellesley

In the Hunnewell Arboretum, home of H. H. Hunnewell.

There seems no reason to doubt that this is No. 3 of the original collection so it is accepted as the clonotype of the new selection of this name. See p. 135.

NEW JERSEY

Ridgewood

In 1938 a large specimen was observed at the former residence of A. F. Fuller on Paramus Road. It measured 18 by 25 feet. The estate is now (1983) the Mount St. Andrews Villa, a Sisters of Charity Old Age Home, but the tree is no longer to be found.

NEW YORK

Beacon

At Tioronda, Matteawan (now re-named Beacon) the former home of General Joseph Howland. This property is now the Craig House Sanatorium owned by the Tioronda Company.

This is tree No. 1 of the original four findlings, and was planted by the presumed finder on his own estate. There are four large trunk-like branches which arise from below ground level. In 1938 it was 11 feet in height and about 35 feet in spread. Now that three of the four clones originally comprising the cultivar 'Sargentii' have been treated as fresh selections (See Article 12 of the *Cultivated Code)* receiving the cultivar names 'Brookline', 'Hunnewell' and 'Wodenethe' this tree strictly becomes the clonotype of the cultivar 'Sargentii' in this new, restricted sense, but del Tredici's suggestion of using 'Tioranda' as the cultivar name has considerable practical advantage (See p. 136). It would release 'Sargentii' for use as a loose, collective name.

On the same property, not far from this original seedling, there is an early graft measuring about 9 feet in height and 23 feet in spread in 1938, according to Dr. Stout (Journ. New York Bot. Gard. 40: 153. 1939).

A tree planted at Wodenethe, home of the late Henry Winthrop Sargent, (No. 2 of the original four) is dead, having been reported so in 1923, but since it "did exist" (See Article 42 of the *Cultivated Code)* it is entitled to be regarded as the clonotype of the cultivar 'Wodenethe'. The point of this is that the S. B. Parsons Nursery Catalogue for 1879 clearly states that their stock had been received from H. W. Sargent of Fishkill-on-Hudson (now renamed Beacon) so there is a great probability that many of the trees in the list are of this clone. See p. 136.

Fishkill

A slightly larger plant, 13 feet high and 25 feet wide in 1938, growing on a property owned by Irvin Berrian at the time, was according to Dr. Stout, (loc.cit.) grafted by a Mr. Gaines of Beacon. Scions were probably obtained from the "Tioronda" hemlock.

Hortontown

At the home of the late Joseph Horton, Hortontown, New York, now the residence of Mr. Joob Veldhuis.

The "Horton" is a venerable tree of the Sargent hemlock type, thought to be spontaneous, although not one of the original four by General Howland. It is located above a small farmhouse on a knoll about 500 feet east of the Eastern State Parkway and less than one mile south of Hortontown, Putnam County. It stands only about five miles in a direct line from the village of Fishkill. Dr. Stout talked to the owner of the property, Joseph Horton in 1937, who said he had known it since about 1872 and that it was then at least one-half as large as it is now (1937). Del Tredici (Arnoldia 40 (3): 202. 1980) has studied this tree and investigated its age and origin. See p. 135. It is the clonotype for the new cultivar 'Horton'. See p. 115.

Long Island

In the ground of the Bayard Cutting Arboretum.

Here there are two enormous specimens. One is a tree 15 feet tall with a spread of 35-40 feet, at first having a single trunk 18 inches in diameter, but soon branching into seven large limbs, each 6 inches diameter. The other tree is 12 feet tall by 35 feet across with a single trunk 2½ feet diameter. (It is most probable that these trees were stem-trained as young plants.)

Oakdale

In the Bayard Cutting Arboretum, Oakdale, Long Island, New York.

Dr. George S. Avery, Jr. (Brooklyn Bot. Gar. Record 5(3): 156. 1949.) reported half a dozen specimens ranging from 6½ to 10 feet in height and from 21 to 36 feet in spread. Dr. Avery stated that they were all seventy-five to one hundred years old, but Sargent hemlock was probably not propagated until 1870, thus indicating the oldest one to be a little less than eighty years old in 1949. The largest of these in 1965 was 11 feet high with a spread of 39 feet.

Oyster Bay
At Planting Fields Arboretum, Planting Fields Road, Oyster Bay, Long Island, New York

There are two magnificent specimens in front of the newly reconstructed Camellia House.

Peekskill
A large tree on the estate of the late Henry Ward Beecher was reported in 1944 by W. Judd. Mr. Judd, then propagator at the Arnold Arboretum, considered it to be another, unrecorded seedling from the original Howland collection. It was 32 feet across in 1944, about the same size as the "Tioronda" hemlock, or perhaps a little larger. Dr. T. R. Dudley points out that no documentation exists to confirm the accuracy of Mr. Judd's opinion, but it would be an interesting research exercise to ascertain whether Beecher's known contact with General Howland establish him as a likely recipient of a plant when the distribution was being made.

Poughkeepie
On the campus of Vasar College.
Almost certainly one of the early Parsons' propagations. It is a tree with three trunks, each 8″ diam., 10 feet high and with a crown diameter of 30 feet.

Troy
Two specimens have been seen by the author about 15 feet high and with a spread of about 25 feet. These two trees were reported by H. T. Ruckaberle of Granville. One is on a private property on Oakwood Avenue, the other is in Oakwood Cemetery.

OHIO

Norwich
In 1940 a specimen was observed by the author at the site of the former Miller Nursery. It was 15 feet high with a 25 foot spread. The trunk measured 25 inches in diameter about 8 inches from the ground. Above that it branched into two main trunks.

PENNSYLVANIA

The following trees are in Philadelphia or its vicinity.

In Fairmount Park
Site of the Centennial Exposition, 1876.
In 1938 the largest measured 15×30 feet. In 1972 the same tree was 18 feet high with an average spread of 36 feet. Diameter of the trunk was 2 feet 9 inches at 1 foot (Fig. 106).

At 5001 Grant Avenue
Three large Sargent hemlocks have recently been found on the grounds of the former Tonner Estate, now a Convention Center for the Lutheran Church of America. They were planted about 8 feet apart and are all low branched. The largest is 16 feet high with an average spread of 37 feet. The diameter of the trunk is 26 inches at 10 inches. These measurements were taken in 1972.

At Morris Arboretum, Meadowbrook Avenue.
Two trees are growing here. In 1938 the larger one was about 20 feet high and about

21 feet across. In 1972 it was 23 feet high with an average spread of 32 feet. The diameter breast high was 20 inches. It probably had been staked for many years.

At Bryn Mawr College

Here there are nine trees whose tops merged according to Stout (Journ. New York Bot. Gard. **40**: 159. 1939). The largest in 1939 was about 12 feet tall and about 18 feet in breadth. In 1972 seven trees were observed along the wall of the older library building. The largest is 14 feet high and 39 feet in spread with trunk diameter of 11½ inches. The largest specimen on the campus is behind Rockefeller Hall, 2 feet 7 inches in diameter at 3 inches from the ground. It is 18 feet high with a spread of 29 feet.

Wayne

At Inver House (on the Darby-paoli Road)

According to Stout (loc. cit.) there were at one time eleven specimens of approximately the same age. These varied from 10 to 12 feet high and from 19 to 26 feet broad. In 1972 the writer visited this property and found ten plants of the original eleven. One of the specimens in the row of seven has died. Two more are quite some distance apart along the main driveway of the mansion. There are two plants bordering the parking area. One of these, near the row of six, is the largest in the group. This is 17 feet high and 36 feet in greatest spread with a trunk diameter of 2 feet at 6 inches. All of the specimens at Inver House are low branched and most of them are to some extent depressed by shade and root competition.

Wynnewood

On the estate of the late Mary K. Gibson (now owned by John W. Merriam), Penn Road above the Wynnewood railroad station.

There were eight specimens here, said to date from the time of the Centennial Exposition. The largest of four remaining plants now measures 15 feet in height and about 20 feet in greatest breadth. The diameter of the trunk is 18 inches at 2 feet.

A small plant that is regarded as a seedling was growing in a shaded location in underbrush on the same property. In 1938 this 4-foot-high plant was transplanted to the Hemlock Arboretum at Far Country. In 1972 it was 14 feet high and about 23 feet wide. All of these hemlocks are low branched. The ones in Wynnewood have been dwarfed by shade and root competition, and the seedling at Far Country, under more favorable conditions, has grown faster.

T. canadensis 'Sargentii'. Three of the four trees planted before the 1876 Centennial in Fairmount Park, Philadelphia, Pennsylvania. Photograph taken in 1974.

One of the specimens of *T. canadensis* 'Sargentii' on the Tonner Estate (now a convention center of the Lutheran Church of America). Philadelphia, Pennsylvania. Photograph taken in 1974.

T. canadensis 'Sargentii' at Wynnewood, Pennsylvania, on an estate now owned by John W. Merriam. One of several planted at the time of the Centennial Exposition.

HEMLOCKS FOR HEDGES

The Canadian hemlock, and still more the Carolina hemlock because of its denser habit and slower growth, makes an excellent hedge, as will be seen in the photographs below and opposite, since they stand clipping well. A hemlock screen makes a striking background for a flower or shrub border. It avoids the sombre colour of the Yew. It forms a beautiful hedge, whether 5 or 20 feet high, and requires clipping only once or twice a year. According to R. W. Oliver (*Hedges for Canadian Gardens* 1973), the Canada hemlock is still a good hedge after sixty years.

Canada hemlock hedge, Middletown, Connecticut.

Canada hemlock hedge, grounds of Lord and Burnham, Irvington, New York.

Sculptured hemlocks, St. Davids, Pennsylvania.

T. caroliniana hedges growing freely. This species responds well to annual clipping and forms a very dense hedge.

THE NAMED GARDEN FORMS

Although it has long been known that the range of variation of which the Canadian hemlock is capable included forms that were so slow-growing that they could safely be treated as "compact" and even "dwarf" and be planted despite the warning regarding space mentioned in a previous paragraph, and that these diminutive forms made excellent garden plants, very little use has been made of them.

In the past, a few nurserymen in this country and abroad offered, from time to time, a limited number of named cultivars of Canada hemlock. These, with a few exceptions, had been originated and selected, or at least named, by Europeans. The only American botanist who was sufficiently interested in hemlock variations to actually name and describe a few new ones was Liberty Hyde Bailey (*The Cultivated Conifers* 123. 1933). Therefore, since many more seedlings are grown in this county than abroad (Canada hemlock does not really grow satisfactorily in most parts of Europe), probably nine tenths of the distinctive variants had remained unrecognized and unknown when the author commenced his study of this genus in 1939.

The typical Canada hemlock is the most popular conifer in the New England and Middle Atlantic states. Why has not that popularity been extended to its cultivars? One reason has been the difficulty of propagation. Being mutations, these variants could not (at least reliably) be reproduced from seed: they had to be propagated vegetatively, and since *Tsuga* were in the "Difficult to impossible" category to root from cuttings, this meant propagation by grafting onto seedlings. This had several drawbacks; it needed special equipment and experienced staff so it was expensive, the success rate was low and liable to reduction from "delayed incompatibility" and also the influence of the understock often more or less destroyed the character of the variant being propagated.

With the advent of root-promoting chemicals (which have proved very efficaceous in the propagation of hemlocks from cuttings) this objection no longer applies. But another hindrance to the wider recognition of their landscape value has been the lack of any authoritative treatise or descriptive literature in the gardening press. It is to be hoped that the present book will overcome this difficulty and open the way to a much wider appreciation amongst landscape architects and the gardening public of the garden value of many of the selected clones.

Only the Sargent hemlock can lay claim to being widely known, distributed, and cultivated. Other weeping cultivars which could likewise be useful as specimens or for accent in border plantings or drooping over pools, are 'Kelsey's Weeping' and (where a miniature plant is required) 'Cole'. The latter should be grown in a well-drained site with pebble mulch and partial shade. Spreading hemlocks could be used in the same way as well as in foundation plantings. The best of these are probably 'Bennett' and 'Curtis Spreader'. Similar to 'Bennett' but with a different origin is 'Minima'. Descriptions of these and the following clones will be found in Chapter Five.

Where globose shapes are desired the best are probably 'Rock Creek Globe' and 'Curtis Ideal', a conical globose plant with good foliage and texture. Most upright cultivars are rather heavy and formal, but there are several in the COMPACT GROUP that are more graceful and full-growing, and they hold up well with age. These are 'Bradshaw', a wide, pyramidal plant, 'Fremdii', a hardy, compact form, and 'Geneva', a bright green upright hemlock with good foliage.

There are also many slow-growing or dwarf cultivars that are more interesting to the hobbyist and collector than to the landscape designer and nurseryman. Their foliage and habit cover a wide range of variability. Some of special merit might include 'Abbott's Pygmy', 'Betty Rose', 'Minuta', 'Brandley', 'Muttontown', 'Stewart's Gem', 'Dwarf Whitetip', 'Essex', 'Everitt Golden', 'Henry Hohman', 'Horsford Dwarf', 'Jervis' and 'Rugg's Washington Dwarf'. Examples of superior free-growing cultivars are 'Dawsoniana' and 'Valentine Yewlike', also 'Westonigra'.

Thus far, only collectors of conifers, alpine gardeners, those who work with Japanese gardens (amateur or professional) and a few landscape architects are at all appreciative of hemlock variations. The late Fred Bergman, proprietor of Raraflora, Feasterville, Pennsylvania, used to claim that the average individual interested in horticultural varieties is enthusiastic about cultivars of pines, spruces, and Hinoki cypress, but passes by hemlocks without comment. But for the collector of dwarf and slow-growing conifers, the hemlocks are the aristocrats.

T. canadensis 'Sargentii' trained on a wire at Raraflora, Feasterville, Pennsylvania.

Part of the collection at the Hemlock Arboretum at Far Country, Philadelphia, Pennsylvania.

3 *THE SPECIES IN CULTIVATION*

The Hemlocks, or Hemlock Spruces form a Genus within the Family PINACEAE comprising ten or more species and one alleged natural hybrid. The synonymies below and in Chapters Four and Six are taken from the *Bibliography of Cultivated Trees and Shrubs* by the late Professor Alfred Rehder, published in 1949.

TSUGA

Carrière, Traité Conif. 185. 1855.—Flous in Trav. Lab. For. Toulouse, II. **4,3**: 136 p., fig. (Rév. Tsuga) 1936; in Bull. Soc. Hist. Nat. Toulouse, **71**: 315-450, fig. 1937.
Pinus Linnaeus, Sp. Pl. 1000. 1753, p. p.
Abies Miller, Gard. Dict. abridg. ed., **1** :1754, p. p.; Gard. Dict. ed. 8, 1768, p. p.—Loudon, Arb. Brit. **4**: 2293. 1838, p. p., quoad sect. II (p. 2319).
Pinus-Abies Weston, Bot. Univ. **1**: 210. 1770, p. p.
Pinus II. *Abies* Münchhausen, Hausvat. **5**: 222. 1770, p. p.
Peuce L. C. Richard in Ann. Mus. Hist. Nat. Paris, **16**: 298. 1810, p. p.
Pinus [group] *Peuce* Sweet, Hort. Brit. ed. 2, 475. 1830, p. p.
Pinus sect. *Abies* D. Don in Lambert, Descr. Pinus, 3: sub *P. Pindrow* [p. 2] 1837, p. p.
Picea sect. *Desciscentes* Link in Linnaea, **15**: 523. 1841, p. p.
Abies sect. *Micropeuce* Spach, Hist. Nat. Vég. Phan. **11**: 424. 1842.
Pinus sect. *Tsuga Micropeuce* (Spach) Endlicher, Syn. Conif. 83. 1847.
Abies sect. II. *Tsuga* Gordon, Pinet. **13**: 1858, p. p. typ.—K. Koch, Dendr. **2, 2**: 248. 1873 "subgen.", p. p. typ.
Pinus sect. *Tsuga* Parlatore in De Candolle, Prodr. **16,2**: 427. 1868, p. p.
Picea subgen. *Tsuga* Bertrand in Ann. Sci. Nat. Bot. sér. 5, **20**: 87. 1874.
Tsuga sect. I. *Eutsuga* [Engelm.] Eichler in Nat. Pflanzenfam. II. **1**: 80. 1889.
LECTOTYPUS: *Pinus canadensis* L.=*T. canadensis* (L.) Carr.

The genus is divided naturally into two sections that have been recognised by botanists:

Sect. 1. MICROPEUCE [Spach] Schneider in Silva Tarouca, Uns. Freil.-Nadelh. 291. 1913.—Fitzpatrick in Sci. Proc. Roy. Dublin Soc. new ser. **19**: 201. 1929.
Tsuga sect. *Eutsuga* Engelmann in S. Watson, Bot. Calif. **2**: 120. 1880.—Beissner, Handb. Nadelh. 394. 1891.

BRANCHLETS not all in one plane. LEAVES spreading all round the shoot, rounded or keeled above, bearing stomata on both surfaces. CONES oblong cylindric, composed of numerous scales. In this Section come *T. mertensiana* and the alleged hybrid *T.* × *Jeffreyi*.

Sect. 2. HESPEROPEUCE Engelmann in S. Watson, Bot. Calif. **2**: 121. 1880.—Beissner, Handb. Nadelh. 394. 1891.
Hesperopeuce (Engelm.) Lemmon in Bienn. Rep. Calif. State Board For. **3**: 126, t. 1890; Cone-bear. Trees Pacific Slope, 12. 1892.

BRANCHLETS all in one plane. LEAVES pectinately arranged, or apparently so, flat, grooved above, bearing stomata on the lower surface only. CONES ovoid, small, composed of few scales. Save as above, all the species in cultivation come in this Section.

All species form monoecious evergreen trees, medium to tall (often shorter in cultivation) with slender spreading branches, irregularly whorled and much ramified, the top-shoot arched and the lateral terminals often gracefully pendulous.

WINTER BUDS: Minute, globose or ovoid, not resinous.

LEAVES: Small, usually blunt, spirally arranged but thrown into a 2-ranked arrangement by the twisting of the leaf-stalks or (Section *Hesperopeuce*) spreading all round the shoot, persistent for several years, closely set on shining, cushion-like projections which are separated by grooves; with a single resin canal situated between the vascular bundle and the lower surface. The stalks are thin, short and pressed against the shoot.

FLOWERS: Solitary, on the previous year's shoot. *Staminate (Male):* Globose, in the axils of the leaves, composed of numerous anthers. *Pistillate (Female):* Terminal on lateral shoots, composed of spirally arranged scales.

CONES: Solitary, small (save only in *T. mertensiana*), ovate or oblong with thin flexible persistent scales, ripening the first year but not falling or breaking up when ripe.

SEEDS: Two on each scale, winged.

The graceful habit, due to the slender, drooping terminal branchlets, with the usually 2-ranked arrangement and the slender leaf-stalks pressed against the shoots are good distinguishing characteristics. The cones do not disintegrate on ripening as they do in the genus *Abies*, and the leaves, on falling, do not leave the peg-like projection characteristic of the genus *Picea*. The late Professor Alfred Rehder, writing in the *Standard Cyclopedia of Horticulture* (3390. 1944.) tells us: "The *Tsuga* are closely allied to *Abies* and *Picea* and differ little in the structure of the flowers; the cones are very similar to those of the larch, but the leaves, which are very much like those of *Abies* in their outward appearance, though smaller, are very different in their internal structure from all allied genera, having a solitary resin-duct situated in the middle of the leaf below the fibro-vascular bundle".

KEY TO SPECIES CULTIVATED IN U.S.A. AND EUROPE*

The following key (See Footnote) covers all the species of *Tsuga* and hybrids known to be in cultivation in North America and Britain. It makes use of foliar characteristics only, so may be used when cones are not available. Other distinguishing characters will be found in the descriptions of Species in Chapter Six.

1. Leaves lying all round shoot; thick; glaucous both surfaces . . . . *mertensiana*
   Leaves pectinate or above shoot only; thin; not glaucous above 2
2. Leaves greenish white banded beneath 3
   Leaves with bright white bands beneath 5
3. Shoot long, sparse-pubescent; leaves very slender, perpendicular, some
      backwards. . . . . *jeffreyi*
   Shoot with short scattered pubescence; leaves broad, pectinate, irregular
      lengths. 4
4. Shoot pink-brown; pubescent on pulvini only, leaf to 2 cm. . . . . *forrestii*
   Shoot creamy white; pubescence scattered all over; leaf to 1.5 cm. . . . . *chinensis*
5. Shoot dark brown or orange. 6
   Shoot white, fawn, buff, cream etc. 7
6. Shoot shining red-brown; leaves distant, slender, not flat two-ranked. . . . . . *caroliniana*
   Shoot bright brown or orange, leaves close, very broad, flat two-ranked . . . *diversifolia*

*The Key to the Species is taken from *Conifers in the British Isles* A. F. Mitchell (British Forestry Commission Booklet No 333. Her Majesty's Stationery Office 1972.) by kind permission of Mr. Mitchell and the Forestry Commission. All material thus used is Crown Copyright.

 7. Shoot quite glabrous, pale buff. . . . . *sieboldii*
  Shoot conspicuously hairy 8
 8. Leaves taper from base or near 9
  Leaves parallel-sided. 10
 9. Leaves hard, distant, well forward on shoot, two broad and brilliant white
   bands beneath . . . . *dumosa*
  Leaves soft, dense except on long shoots, a line lying reversed along
   and close to shoot, two narrow white bands beneath . . . . *canadensis*
10. Leaf deep blackish-green after few weeks; two broad blue-white bands
   beneath; densely set, short mid-shoot leaves irregular, upstanding . . . . . . *heterophylla*
  Leaf pale shining green; two very broad, very white bands beneath;
   rather sparse, no upstanding leaves on mid-line . . . . *yunnanensis*

Professor Dr. P. Schutt and A. John, both of Forstbotanisches Institut in Munich (Munchen) are working on a scheme for the identification of conifer species by microscopical examination of the leaves. The first part of their thesis, "Blattan-atomische Merkmale als Hiftsmittel fur die Artdiagnose von Nadelbaumen. 1. Tsuga-Arten." (Mitt. d. d. d. Ges. **70**: 103. 1978) deals with *Tsuga* species, but unfortunately for some readers, it is in German.

OTHER SPECIES

A number of other specific names will be found in the literature. These are usually synonyms, little or imperfectly known species or names now transferred to a different genus. Descriptions are taken from *Manual of Ornamental Conifers* Den Ouden and Boom, 1965 by permission of Dr. B. K. Boom.

T. ajanensis (Lindley and Gordon) Regel. A synonym of *Picea jezoensis* (Sieb. and Zucc.) Carr.

T. albertiana Sénéclauze. A synonym of *T. heterophylla* (Raf.) Sargent.

T. araragi (Sieb.) Koehne. A synonym of *T. sieboldii* Carr.

T. blaringhemii Flous in Bull. Soc. Hist. Nat. Toulouse **69**: 2. 1936.
Closely related to *T. diversifolia,* differing; branchlets dark chestnut-brown, densely pubescent; winter-buds less in number; scales oblong, margins finely toothed; rows of stomata narrower; cones globose, 2 cm diameter; scales nearly triangular, with 2 small ears at the base; seed-wing to 6 mm long.
Central Hondo (Japan).

T. brunoniana (Wallich) Carr. A synonym of *T. dumosa.* (D. Don) Eichler.

T. calcarea Downie in Not. Roy. Bot. Gard. Edinb. **14**: 17. 1923.
Differs from *T. chinensis* by its reddish-brown branchlets, the shorter leaves (6-10 cm), the finely toothed margin and the erect cone-scales.
W. China (Yunnan) at high elevations.

T. crassifolia Flous in Bull. Soc. Hist. Nat. Toulouse **69**: 4. 1936.
Tree, in habit and size similar to *T. blaringhemii;* branchlets short-pubescent when young, light orange, later silvery grey; winter-buds conical, 5 mm long, 2 mm broad; scales very narrow, strongly pubescent; leaves more or less curved, entire, 12-25 mm long, 1-1.3 mm broad, apex blunt, distinctly narrowed into an about 2 mm long stalk, convex above, shallowly keeled and with 4-6 lines of stomata on both sides of the midrib; cones ovoid-cylindrical, 3.5-6 cm long, 2-2.5 cm broad, violet when young, light brown when mature; scales about as long as broad, nearly round and auricled; bracts

rather long, reaching half the scale, toothed, apex pointed; seeds 4 mm long, with 3-4 large glands; wing 7 mm long, apex rounded.
California and Nevada (U.S.A.).

T. douglasii Carr. A synonym of *Pseudotsuga menziesii* (Mirbel) Franco.

T. dura Downie in Not. Roy. Bot. Gard. Edinb. **14**: 16. 1923.
Differs from *T. yunnanensis* Mast. by the more greyish-brown branchlets and the thicker more woody and not striped cone-scales.
Yunnan Province, W. China.

T. formosana Hay. in Gard. Chron. **1908,1**: 194. Differs from *T. sieboldii* by its obtuse bud-scales, its shorter (-13 mm) leaves which are not notched at apex and its smaller cones. Found on Mount Morrison, Taiwan by S. Nagasawa in 1905.

T. forrestii Downie in Not. Roy. Bot. Edinb. **14**: 18. 1923.
Differs from *T. chinensis* by its yellowish branchlets, its larger leaves (to 25 mm long) and its larger, short-stalked cones.
Yunnan Province, W. China in high elevations.

T. hanburyana Name only. A synonym of *T. sieboldii.*

T. hookeriana Carr. A synonym of *T. mertensiana* (Bongard) Carr.

T. intermedia Hand.-Mazz. in Anz. Akad. Wiss. Wien, Math. Nat. Kl. **61**: 82. 1924.
Tree to 40 m tall; branchlets light brown when young, older grey, densely bristly pubescent; leaves 10-20 mm long, 1.5-2 mm broad, apex rounded or slightly notched, entire, thick, margins slightly recurved, dark olive green with some rows of stomata above and with 10-14 rows of stomata beneath; cones sessile, ovoid, 13-20 mm long and broad, thick, moderately pointed; sterile scales 4-5 mm long, notched, fertile scales roundish, narrowed at base, 7-9 mm across, woody, broad and thin margined, slightly shining, very finely streaked; bracts appressed; seeds including the ovoid wing, 6-6.5 mm long.
It is intermediate in all characters between *T. yunnanensis* and *T. chinensis* and it occurs in rain forests at an elevation of 2500-3000 m in Yunnan Province on the Burma-Tibetan boundery.

T. leptophylla Hand.-Mazz. in Anz. Ak. Wiss. Wien, Math. Nat. Kl., **61**: 83. 1924.
Closely related to *T. dumosa,* but differs from the latter by its glabrescent branches and the entire leaves.
Yunnan Province, W. China at high elevations.

T. lindleyana Roezl. A synonym of *Pseudotsuga menziesii* (Mirbel) Franco.

T. longibracteata Cheng in Contrib. Biol. Lab. Sci. Soc. China 7: 1. 1932.
Tree to 10 m tall; trunk 0.25 m in diameter; branchlets yellowish-brown to brown, older becoming grey, glabrous or scattered-pubescent; winter-buds ovoid-pointed, 2-4 mm long, ± 2 mm broad; scales closely appressed, ovoid, shining brown, margins slightly fimbriate; leaves 1.5 mm long-stalked, contorted, linear, 11-22 mm long, 1-2 mm broad, entire, apex pointed, light green beneath with rows of stomata on both sides; cones short-stalked, ovoid to oblong-ovoid, 2-3 cm long, 1.8-2 cm broad; scales almost rhomboidical, 1-1.5 cm long, 1-1.3 cm broad, auricled, streaked and slightly pubescent; bracts subspatulate, ± 8 mm long, only half deflexed, lower part narrow, upper part broad, apex pointed, slightly toothed at the upper-margin; seeds 4 mm long, glands large; wing 6 mm long.
Kweichow Province, China.
It differs very much from both the other Chinese species. *T. chinensis* has the branchlets yellow and pubescent in the grooves, leaves notched at the apex, with stomata on both surfaces, cones sessile and bracts not exserted. *T. yunnannensis* has pubescent

branchlets, leaves blunt at the apex, with stomata on one surface, distinct bluish-green beneath, and small cones with hidden bracts.

T. patens Downie in Not. Roy. Bot. Gard. Edinb. **14**: 16. 1923.
Differs from *T. chinensis* by the larger winter-buds (3 mm long, 2 mm broad), the rounded bud-scales, and the mostly finely toothed leaves.
Hupei Province, W. China. Found in 1907 by E. H. Wilson.

T. pattoniana (Murr.) Parl. in D.C. A synonym of *T. mertensiana* (Bongard) Carr.

T. stairii Now *Pseudotsuga menziesii* 'Stairii'.

T. taxifolia Kuntze. Now *Pseudotsuga menziesii.* (Mirbel) Franco.

T. tsuja Murray. A synonym of *T. sieboldii* Carr.

T. wardii Downie in Not. Roy. Bot. Gard. Edinb. **14**: 17. 1923.
Differs from *T. chinensis* by the leaves which are narrow, finely toothed and white beneath.
Yunnan Province, W. China at high elevations. Found by Ward in 1913.

T. williamsonii Vos. A synonym of *T. mertensiana* (Bongard) Carr.

CONFUSION BETWEEN THE SPECIES

There has been in the past and probably still is confusion in the nursery trade between even the most widely met-with species, but despite a rather close similarity between all hemlocks this should not be necessary.

It should be easy to distinguish the two eastern American species by foliar characters alone. *T. canadensis* has its short leaves on the upper side of the branchlets pointing forwards along the shoot but twisted so as to shew the white underside. *T. caroliniana* has similar leaves also held forward but not twisted.

The two western American species are also sometimes confused, but this is quite unjustified since even a novice should be able to tell them apart. The Western hemlock, *T. heterophylla* holds its thin, flat leaves in the flat, spreading arrangement characteristic of hemlocks and the leaves have stomata on one side only. The Mountain hemlock, *T. mertensiana* holds its thick, blue-green leaves forward and all round the shoot, and the leaves have stomata on both sides.

Confusion between the two Japanese species *T. sieboldii* and *T. diversifolia* is perhaps more easy to understand, since nurserymen in this country are dependent upon foreign seed. The foliage of *T. diversifolia* is very dense and regular and the dark green leaves are conspicuously white beneath (hence the vernacular name "Rice Tree") and the brightly coloured shoots are finely pubescent. *T. sieboldii* has leaves that are longer and more loosely held and have less conspicuous stomatic bands.

Tsuga canadensis

4 *TSUGA CANADENSIS, THE CANADIAN HEMLOCK*

(See also Chapter Five for cultivars.)

TSUGA CANADENSIS (L.) CARRIÈRE, CANADIAN HEMLOCK; EASTERN HEMLOCK*

Tsuga canadensis (L.) Carrière, Traité Conif. 189. 1855.—Sargent, Silva N. Am. **12**: 63, t.603. 1898.
Pinus canadensis Linnaeus, Sp. Pl. ed. 3, **2**: 1421. 1763.
Abies americana Miller, Dict. Gard. ed. 8, *A.* no. 6. 1768.
Pinus-Abies americana Weston, Bot. Univ. **1**: 211. 1770.—Marshall, Arbust. Am. 103. 1785.
Pinus Abies canadensis Muenchhausen, Hausvat. **5**: 223. 1770.
Pinus americana Du Roi, Obs. Bot. 39. 1771.
Pinus mariana Gaertner, Fruct. Sem. **2**: 59, t. 9l, fig. a-g, 1791, exclud. syn. Seligm., Mill.; non Du Roi. 1771.
Pinus pendula Salisbury, Prodr. Stirp. Chap. Allert. 399. 1796, non Aiton. 1789.
Abies canadensis Michaux, Fl. Bor.-Am. **2**: 206. 1803, non Miller. 1768.
Abies pectinata Poiret, Encycl. Méth. Bot. **6**: 523. 1804, non Gilibert. 1792.
Picea canadensis (L.) Link in Linnaea, **15**: 524. 1841.
Tsuga americana (Mill.) Farwell in Bull. Torrey Bot. Club. **41**: 629. 1915.

A tree usually 60-70 feet, and occasionally 100 feet high, with long, slender horizontal lateral branches with pendulous branchlets, forming a broad pyramidal or gradually tapering head. An important, widespread forest tree. Canada from East of the Rocky Mountains to Nova Scotia; U.S.A. Lake States and Appalachian Mountains south to Alabama. It was introduced in 1736 to Britain, where it is common in arboreta and (mostly only the cultivars) in gardens.

BARK. Young trees bright orange-brown with purple-brown scaly ridges. Old trees dark purplish grey with a lattice-work of thick scaly ridges.

CROWN. Young trees broadly conic with obtuse apex and arching slender shoots. Old trees irregularly broad-conic, seldom with a single bole for more than a few feet.

FOLIAGE. Shoot pale brown or cream, with a dense, curly, pale orange pubescence. Second year dark grey or grey-brown.

BUD. Hidden by leaves, broad conic, red-brown.

LEAVES. Pectinate and slightly depressed beneath; above over the shoot and forwards, mid line leaves showing white undersides; broadest at base, tapering slightly to rounded tip. Underside with broad green margins, two white bands and a green midrib which a lens shows to be white centred with a narrow green line each side. Crushed foliage has a sweet lemony scent.

CONE. Ovoid, pendulous, usually numerous, 1.5-2.5 cm, grey-brown; scales rounded, obscurely toothed.

GROWTH. Despite its very slow growth at the seedling stage it grows rapidly, when established, with shoots to 2 feet or more long, but growth in height slows rapidly as the leading shoot tends to lose dominance. There are very few dated trees to show the increase in girth on a single bole but remeasurements show only one approaching one

---

*The descriptions here and in Chapter Six are based on and the line drawings taken from *Conifers in the British Isles*, A. F. Mitchell (British Forestry Booklet No. 333. Her Majesty's Stationery Office 1972.) by kind permission of Mr. Mitchell and the Forestry Commission. All material thus used is Crown Copyright.

inch per year and the majority less than half this rate. Two dated trees in Britain have, however, exceeded one inch per year.

RECOGNITION. The dark, heavily ridged bark and broad crown at once distinguish this from *T. heterophylla*. The foliage is distinguished from all other *Tsugas* by the line of leaves upside-down along the line of the shoot.

An outstanding feature of this species is its great variability from seed and its proneness to develop mutations of the witches' broom type. These occur too sporadically to classify as infraspecific botanical entities, but a large number of the variants (many of them dwarf or compact in growth) make excellent garden plants. Details of the clones that have received cultivar names will be found in Chapter Five.

Several testimonies to the grace and beauty of the Canadian hemlock have already appeared in Chapter Two, so here it must suffice to add the following tribute.

One of our earliest horticultural writers, Thomas Meehan who, on the title-page of his *The American Handbook of Ornamental Trees* published in Philadelphia in 1853, describes himself simply as 'Gardener,' writes of this species (which, of course, in those days he called *Abies canadensis*):

> It would not be exaggeration to pronounce this the most beautiful evergreen in cultivation. Beautiful as many of the new pines are, few approach this. It has regularity without formality; and, in any point of view, elegance and gracefulness. Its habit is frequently so erect as to approach the fastigiate; yet the ends of its branches are as pendulous as a Babylonian willow. Its color is not of that mournful cast so common to other Pinaceae; nor of that consumptive-looking hue so connected with sickliness. Stepping between these it is suggestive of innocence and lightness, which cannot fail to attract admirers for it, in whatever situation placed. It will make the prettiest object when grown by itself; but it is a tree that has no aversion to company.

The early lumbermen were convinced that there were two distinct forms of this species which they distinguished as "red" and "white" hemlocks. Some even believed there were two species. These would often occur sporadically, but sometimes in pure stands, "red" hemlocks often on high ground or in difficult country whereas "white" trees preferred the richer soil in the valleys. The difference in habit, shape of crown and texture of bark allegedly made the distinction clear to an experienced forester, but cases of doubt could be resolved by peeling a strip of bark and noting the colour on the back. The same distinction was said to appear in the colour of the heartwood. The timber of "white" hemlock was held to be much superior.

The distinction has, however, never been accepted either by botanists or by authorities in the forestry world. The differences are related to the age and vigour of the tree, especially in relation to its environment and nutritional status. It is characteristic of mature trees to develop a broader crown and coarse bark, and the timber from prosperous trees in rich lowland soil would clearly be sounder than that from much older trees of the same size in a difficult, exposed situation where ring shakes would be expected as a result of wind action.

PHYSIOLOGICAL CHARACTERS

HABIT OF GROWTH

The normal Canadian Hemlock of the forest is at first conical, but in maturity is a broad-crowned tree reaching to 20 m (occasionally to 30 m) with a trunk 1 m or more in diameter, unless forked near the ground, which occurs. It has a very wide natural distribution and was a characteristic tree of the virgin forests of North-east America. There it is to be found in rocky, barren spots but seems happiest on north-facing slopes and in steep valleys where it is in partial shade and finds plenty of water at the root. This affords a valuable clue to the conditions in which it will be successful in cultivation. Because of its small leaves and the thin, usually pendulous branchlets it is a very graceful tree.

There is an interesting tradition that the leader in trees growing in the open always arches over in the same direction. William H. Harlow, a forestry Professor at the New York State College of Forestry at Syracuse University, wrote that "the terminal shoot in contrast to that of other eastern conifers (Carolina hemlock excepted) is flexible and tends to curve away from the direction of the prevailing winds. This identifies hemlock from a considerable distance and may also be used, with reservations, as a 'natural compass' since the slender tips commonly lean toward the northeast" (Harlow and Harrar, *Textbook of Dendrology,* 153. 1937.). On this matter Mr. Jenkins wrote (*Heml. Arb. Bull.* 37. 1942.), as follows:

> My old friend and fellow hemlock enthusiast, Frank L. Abbott of Worcester, Massachusetts, told me some time ago that the terminal shoots of hemlock, *T. canadensis,* always point in a certain easterly direction, a few points south of east. He writes (November 3, 1941), 'Have just returned from a hunting trip to Forest City, Maine. In that section, as well in Vermont, all Canada hemlocks that have a normal top growth and have matured to approximately 20 feet, have the terminal growth pointing a few points south of east. This would be an important fact to know when one is lost in the big woods. There are some exceptions to this rule but over 90 percent of Canada hemlock will be found growing in this manner'.

Mr. Jenkins continued this discussion in *Heml. Arb. Bull.* 38. 1942:

> It is the history of human inventions to think something new has been discovered when in many cases it has long been known. The little item in the last bulletin about using hemlocks as a compass when lost in the woods has brought a number of confirming reports. It seems to be an established item of forest lore.

VARIATION IN HABIT OF GROWTH

As might be expected in a species so variable in almost every respect, there is considerable variation in habit and a great diversity is displayed by different variations of hemlock. There are many departures from the typical in this respect.

All the young main shoots, including the central leader are thin and at first very flexible and consequently pendulous, this being one of the characteristics that make the hemlocks so attractive. But as the year goes on and the branches lignify the main branches stiffen and the leader becomes erect. This habit is not by any means confined to hemlocks—it can be observed in other genera, examples commonly met with being the Himalayan cedar, *Cedrus deodara* and some species of *Cupressus* and *Chamaecyparis.*

In simple, gardeners' language, the leader at first "nods" but by Fall "pulls itself up straight". The result is a steady development year after year of a straight trunk and the normal pattern of stiff and (usually) ascending main branches. In the pines this process is reversed. The young shoots, usually suggestively referred to as "candles" are at first stiffly erect, but come winter they have lowered themselves to their final angle (which in the case of the prostrate cultivers is horizontal, or nearly so.)

But the adage "The child is father to the man" holds as true for woody trees as for *Homo sapiens.* Abnormal habit forms in mature trees can always be traced to some deviation from the normal ability to "pull itself up" that has been with the tree from its earliest days. A sapling that can pull its leader into a vertical position (thus ensuring a straight central trunk in due course) but is unable to stiffen its branches normally will become a *true* pendulous tree. If, in another case the laterals are successful in pulling themselves up to an unusually steep angle and if, despite this the leader can maintain its dominance the little plant will develop into a fastigiate tree. If it cannot, the result will be a multi-stemmed tree.

Cases arise in which the young aspirant finds itself unable to pull its branches, any of them, up beyond an angle of about 45° above the horizontal. Try how it may, this is all it can ever manage; its one, two or three little branches finish up at this angle. Many years later it will be struggling with the same limitation, but in the meantime it will have become a magnificent specimen of the Sargent hemlock. It will be more exactly a large bush than a truly pendulous tree.

The same dependence upon its own genetic make-up puts the young plant at the mercy of the nurseryman responsible for it during its early, formative years. When the landscape potential of the Sargent hemlock was first recognised its propensity to increase in width rather than height was regarded as a disadvantage. Attempts at "top working" were unsatisfactory, according to a report quoted by Stout (*Journ. New York Bot. Gar.* **40**: 158. 1939):

> It may be readily grafted on high stocks, but it does not thrive as well, the naked stem cracks and suffers, and the massive foliage, like that of most evergreens perched upon high stems, is too heavy for grace and proportion, and is beaten and tossed by the winds.

The other professional device used was the selection and staking of one of the ascending branches in a vertical position to form a leader. By way of digression, but to show how far this can be taken, the case of the well-known Pfitzer juniper can be cited here. *Juniperus* × *media* 'Pfitzerana' when young is a spreading shrub, structurally consisting of a few main branches rising at about 45°. In time it develops into a low, spreading, open-centred tree with a few large, irregularly spaced branches carrying a thatch of foliage high enough to walk under. The similarity to the Sargent hemlock is striking, but Beissner (*Nadelgehölze* ed. 2, 605-6. 1909) described and gave an illustration of the mother plant as a broad pyramid, 4 m high. This interesting mystery is resolved by another illustration of the same tree, a photograph in the German paper "*Gartenwelt*" (p. 403. 1901) which clearly shows the bamboo to which the leader has been tied still protruding above the tip of the slender leader. Little wonder, then, that so many of the early propagations of the Sargent hemlock retain evidence of this attention in the nursery whence they came.

Finally must come mention of those few plants unable to "pull themselves up" at all and so are destined to remain prostrate plants throughout life.

LEAF RETENTION

For a plant to be an evergreen at all the leaves of one season must remain at least until those of the following season have developed. In the case of the hemlocks the leaves are retained for several seasons. Varying and often conflicting accounts of leaf longevity are given by different authorities for the different species (*T. canadensis:* 1½-3 years; *T. caroliniana:* 2-3 years; *T. diversifolia:* 5-8 years; *T. heterophylla:* 3-4 years; *T. sieboldii:* 4-5 years.)

These discrepant reports are probably due to the fact that leaf persistence is very much influenced by exposure; trees in an open, wind-swept situation lose their leaves very much sooner than those in a sheltered spot. Thomas Meehan (*Heml. Arb. Bull.* 44. 1943) remarks that the density of foliage that is characteristic of *T. caroliniana* is partly due to its stiff and therefore more wind-resistant branches and partly due to the development of leaf-bearing spurs along the whole length of the branch.

RATE OF GROWTH

Everyone acquainted with Canada hemlock knows that it is extremely slow-growing from seed. Jenkins, in *Hemlock Arboretum Bulletin* (22. 1938) printed the following Table which he had compiled from reports sent to him by several leading nurserymen.

Type of plant	Years from seed	Height (Inches)
Seedling	1	1-2
do	2	2-3½
Transplant	3	3-6
do.	4	4-8
do.	5	9-15
do.	6	12-20
do.	7	20-25
do.	8	24-42
do.	9	36-52
do.	10	52-72

Many plants will, of course, have been permanently planted long before ten years from seed. After being set out in their final home, hemlocks will follow the usual pattern, taking a year or two to establish, but thereafter will grow quite rapidly, an annual increase in height of one foot or more being quite usual. Because of the known dependence of the Genus upon adequate access to water, the period before vigorous growth commences may largely depend upon this factor and the ability of the plant to reach underground supply. This amount of annual increase relates to trees of normal vigour; in the case of the compact, dwarf and pygmy garden cultivars it will always be much less, the diminutive cultivars such as 'Minuta' only putting on about half an inch annually.

FLOWERS AND CONES

The male and female flowers are borne in separate clusters on the same tree. The male (pollen-bearing) strobili appear from late April to early June, according to locality, in the axils of the previous year's shoots and consist of rounded yellowish clusters of spirally-arranged, 2-celled anthers. The shorter, female (cone-producing) strobili develop terminally on the lateral shoots of the previous year; they are small, greenish, with rounded scales which are slightly longer than the bracts. The small (6-10 mm) solitary cones on slender stalks mature the same year and are composed of few, overlapping, striated scales, beneath each of which two small winged seeds are borne. The cones ripen in August and by September they have opened and shed the seeds, although the cones remain on the tree until the following year.

SEED AND SEED PRODUCTION

The cones of Canadian hemlock are the smallest of any species of *Tsuga*. The seeds are about 1.6 mm long with a slightly longer terminal wing smaller than either *T. caroliniana* and *T. mertensiana*, but slightly larger than *T. heterophylla*. The seeds are ripening as the cones change from yellow-green to purple-brown in colour and start to fall as the cones begin to dry to a deeper brown. Most seeds fall near the mother tree due to the small wings.

Canadian hemlock produces a crop of seeds annually, varying widely in amount from year to year. Bumper crops occur quite unpredictably, the weather conditions after flowering being apparently critical. Young trees may commence coning at 20-30 years and mature trees continue to produce excellent crops to a great age. Viability is always very low, commonly less than 15%, since only the seeds in the centre part of the cone are fertile. In self-sown seeds the chilling period necessary to complete dormancy is met during the winter, but collected seed needs a period of stratification at low temperature.

MORPHOLOGICAL CHARACTERISTICS

Full descriptions of the clones referred to in the following sections will be found in the alphabetical list in Chapter Five.

BUDS

Little attention has been give by previous writers to the buds, perhaps because they are so small that variation is not very obvious. The color varies a little, but apparently there are only two striking kinds of variable bud characters. One is the amount of pubescence. The buds of typical hemlock are covered with short hairs, denoted as puberulent. But clones in the CINNAMOMEA GROUP have buds that are definitely pubescent in striking contrast to the buds of other specimens at the same stage of maturation.

The other variation is connected with bud production. The excessive production of buds is one expression of the so-called witches' broom habit of growth, and is found particularly in the variants treated under the TWIGGY GROUP. In one instance the author counted eleven minute buds in a nearly circular formation which measured not more than 3 mm in diameter.

The production of adventitious buds on the trunk and branches of hemlock has been observed in at least three instances. Thomas B. Harbeson of Milroy, Pennsylvania, pointed out one striking example. This is a small tree near the breast of Whipple Dam, Center County, Pennsylvania. Adventitious growth which does not live for more than two or three years has been produced on the trunk of this tree for many years. This has been deduced from the presence of numerous bunches of dead branchlets. There is also evidence of this same kind of growth on the trunk and branches of 'Hussii' and 'Greenwood Lake'.

BRANCHLETS

The wide variations in the habit in developing and mature trees can always be traced in the arrangement of the branchlets in the young spray. This must be so, since one is the direct outcome of the other. So, for instance, fastigiate plants will be found to hold their side twigs at a narrow angle, globose and widespreading plants show little or no leader dominance. This enables at least the probable habit of the plant to be deduced from a herbarium specimen.

There are, however, exceptions to this pattern: clones in which the twig arrangement can only be described as irregular or haphazard or (in the case of the curiosity 'Horsford Contorted', which literally forms loops resembling a pig's tail as the twig elongates) as grotesque.

The ultimate branches of most of the plants treated under the TWIGGY GROUP have no definite terminals. A larger branch of such slow-growing plants consists of a number of ultimate branches and still be no longer than a herbarium specimen. Such a branch is often narrow in outline. 'Hussii' is an example; it may be the result of crowding.

Considerable variation has been noted in the length and degree of pubescence of the branchlets. According to Sargent (*Silva of North America* 64. 1898) three-year wood is glabrous. This is not always true. Neither is it always true that the twigs are merely pubescent. The stronger term "tomentose" might almost be applied to several variants such as 'Brandley' in the COMPACT GROUP, 'Von Helms' and 'Redding' in the DENSE-LEAF GROUP (see p. XXX), and in 'Angustifolia'.

Most of the CINNAMON-TIP GROUP have noticeable swellings near the ends of the twigs. These swellings are conspicuous because they are thickly covered with long brown hairs. The most prominent swellings can be observed in that curious little oddity 'Horsford Contorted', where the swellings are often fused in the curve of the branches. In 'Van Dyne' and 'Rugg's Washington Dwarf' these densely-set long brown hairs are absent.

The important characteristic expressed by the words "coarseness" or "fineness" was compared by recording the diameter at the base of the current season's growth. This was a convenient method, because plants with thin twigs will have relatively thin branches.

The range for typical Canada hemlock is apparently 1.2-1.5 mm in diameter. This range includes the majority of the variants, but some fall outside these limits. Those with a range of 0.8-1.1 mm include most of the variants classified under SPREADING and SPARSE-LEAF GROUPS. Plants with unusually thick twigs include some variants treated under the COMPACT and DENSE-LEAF GROUPS.

LEAF ARRANGEMENT

Although, as in other species in the genus, the leaves appear in a spiral arrangement along the shoot, the petioles are bent (i.e. not twisted: the base of the leaf forms an acute angle with the petiole) so that the leaves lie in a two-ranked comb-like arrangement. Flous (*Revision du Genre Tsuga* 60. 1936.) used the bending of the petiole as a key character for *T. canadensis,* since she specified that only in this species does the base of the leaf form an acute angle with the petiole. The author has not found this character very constant. On the other hand, few writers mention that the third row of tiny leaves pointing forward on the top of the shoot are usually reversed so that their white undersides are clearly seen, although the author has found this to be a valuable means of identifying this species. Even most of the peculiar variants which at first glance do not appear to be even close relatives of *T. canadensis* display this character, a rare exception being the cultivar 'Jacqueline Verkade'.

This definite reversal of leaves has been observed in only one other species namely *T. heterophylla* (Western hemlock), and here the tendency is not nearly as pronounced. Any confusion between these two species may be easily settled by examining the white bands on the undersides of the leaves. These are well defined with a distinct green margin in *T. canadensis,* but ill-defined and without a distinct green margin in *T. heterophylla.* With normal vision the observer can distinguish this character without the aid of hand lens.

Numerous deviations from this arrangement occur, but being fairly constant throughout a particular plant, these deviations afford a useful help in the identification of some of the garden forms. Beissner (*Nadelgehölze* 402. 1891) in describing 'Sparsifolia' indicated that the leaves were irregularly disposed about the branchlets. Sénéclauze (*Conifères* 19, 1868) stated that the leaves of 'Latifolia' were not exactly two-ranked. Later writers do not mention leaf arrangement.

Other variations occur. Actually, leaf arrangement is very diverse, and supplies a valuable diagnostic character by which to distinguish between somewhat similar clones. Some cultivars are mostly two-ranked above but quite irregularly disposed below the plane of the branchlets. Examples of this are 'Von Helms', 'Hussii', and

'Bergman's Spreading'. In these cultivars many of the leaves are curved, especially on 'Hussii', some pointing forward, some backward.

A few cultivars have at least a tendency to carry foliage radially. The radial condition does not exist in all branchlets of any individual plant. The leaves of occasional branchlets of a dwarf DENSE-LEAF variation labelled 'Pumila' in the Durand Eastman Park, Rochester, New York, are almost perfectly radial. This is also true of 'Mutton-town'. In most cultivars, the leaf mass points slightly forward. In 'Lewis', the leaves below the plane of the branchlets point forward at an average angle of 30 degrees. A quite unusual arrangement of leaves is found in 'Jervis' wherein the terminals of the branchlets are quite radial and crowded giving a crested or cristate effect. A similar condition can be noted in 'Bacon Cristate.'

LEAF DIMENSIONS

There exists a great difference of opinion concerning the length of leaves of *T. canadensis*. Since this varies so greatly, even along a single shoot, writers in the past have always given two dimensions, these presumably being intended to indicate the range from the shortest to the longest. This range runs from 12 to 19 mm according to Britton and Brown (*Illustrated Flora of the Northern United States, etc.* 56. 1896) and from 8 to 18 mm according to Rehder (*Manual of Cultivated Trees and Shrubs* 39. 1927). Fitschen (*Handbuch der Nadelgehölzkunde* ed 3, 70. 1930) gave the minimum length of 5 mm and Flous (*Revision du Genre Tsuga* 60. 1936) stated that the short leaves of the third row on the upper face are shorter than the lateral leaves, but a minimum length of 8 mm is inaccurate; 3-4 mm is more accurate.

A great deal of statistical work on this subject was done by the author. Included in the study were twenty-two specimens of forest hemlock; that is, trees with the characteristics of typical hemlock. To explore the full range of variability seventy specimens of different cultivars were also studied. These reflected the extremes of variability. The minimum length is 4 mm (occasionally 3 mm) and the maximum length occasionally reaches 16 mm. Therefore 4-16 mm can be considered as the extreme range of leaf length and 7-9 mm, a fair assessment of normal.

LEAF DENSITY

In forest hemlocks the number of leaves for each centimeter of branchlet ranges from 11.3 to 19.5, cultivars range from 7.9 to 27.7. No correlation was found between leaf length and leaf density.

LEAF ANATOMY

Büsgen and Münch (*The Structure and Life of Forest Trees* Trans. by Thos. Thomson. 1929) have written a rather complete discussion on the sun and shade leaves of silver fir. They observed that the shaded leaves are two-ranked, with the leaves on the upper side of the twig very short so that they shade each other as little as possible. In contrast, the leaves exposed to full sun are thicker and longer, with their tips directed vertically upward: "The structure and arrangement of the shade needles allows of the fullest possible utilization of the small amount of light falling on them and the arrangement of the light (or sun) needles insures that the excessive zenith light, rich in ultra-violet rays only grazes their surface." Since hemlock is just as sensitive to light as silver fir, one is not surprised to find that a similar physiological situation exists in the leaves of hemlock. Professor Dr. P. Schutt and A. John (Mitt.d.d.d. Ges. **70**: 103-114. 1978) in the first article—devoted to *Tsuga* species—of a series entitled *Blattanatomische Merkmale als Hilfsmittel fur die Artdiagnose von Nadelbaume*—have presented an interesting study on the determination of *Tsuga* species by a microscopic examination of the anatomy of the leaves. An interesting question is whether there are peculiarities reliably constant within individual plants (and therefore within each clone) which could be of value in distinguishing between similar cultivars.

LEAF SHAPE

The leaves of typical hemlock are nearly linear, broadest somewhat below the middle. Some exceptions may be noted: 'Taxifolia' and 'Jennings Yewlike' have leaves which are broadest very close to the base. Typical leaves may be nearly linear or they may noticeably narrow toward the apex. Perfectly linear leaves with a nearly constant width from the base to the apex are very rare. Most leaves of 'Angustifolia' show this condition although many of the longer leaves are slightly broadest above the middle. Many of the plants in the CINNAMON-TIP GROUP have many leaves that are broadest above the middle. Some are quite broad near the apex and also curve under slightly, suggesting a spoon-like, nearly spatulate shape.

Only a few authorities have noted the breadth and thickness of hemlock leaves. According to Sargent (64. 1905), they average about 1.6 mm in breadth; Fitschen (70. 1930) indicated a maximum breadth of 1.5 mm; Flous (60. 1936) set a maximum breadth of 2 mm. The author, after studying many specimens considers that the actual range of breadth is 1-2 mm. The ANGUSTIFOLIA and CINNAMON-TIP GROUPS usually have abnormally narrow leaves that are somewhat shorter than typical. For thickness of leaves Flous is apparently the only authority who had indicated a definite measurement—0.5 mm. On the other hand, measurements of the dried leaves of many specimens lead the author to conclude that the average thickness is less than 0.5 mm, being nearer 0.25 mm, with a range of 0.15-0.40 mm.

LEAF MARGIN AND APEX

There is considerable disagreement amongst writers as to the extent of the serrulate margin. It is a variable characteristic. The range extends from at least three-quarters of the margin to only a few teeth near the apex. The apex of the leaf is also quite variable in shape, and so is variously described. All authors agree that the apex is rounded or notched. From close observation the author can state that the apex is typically rounded, rarely emarginate, and occasionally acutish, especially on the younger growth.

The various kinds and degrees of serrulation are illustrated on page 52. Typical Canada hemlock has some serrulation, especially near the tip of the leaf. Again the CINNAMON-TIP GROUP shows aberrations. Some leaves are entirely without noticeable teeth, although most leaves have one or two minute teeth near the apex. Two clones related to other groups and also without teeth are illustrated.

The leaves of typical hemlock are mostly obtuse, but even on the same branch, some of the younger leaves may be quite acute and some of the older leaves may be slightly emarginate or notched. In comparing the apices of different clones, what is characteristic for each variant must be considered. Clones in the CINNAMON-TIP GROUP characteristically have a high proportion of acute leaves.

The emarginate condition is most evident on the WIDE-LEAF variations such as 'Curtis Wideleaf.' These have different origins but appear to be identical. The majority of leaves on these specimens are blunt. Occasional emarginate leaves have been observed on several other variations that have not been named.

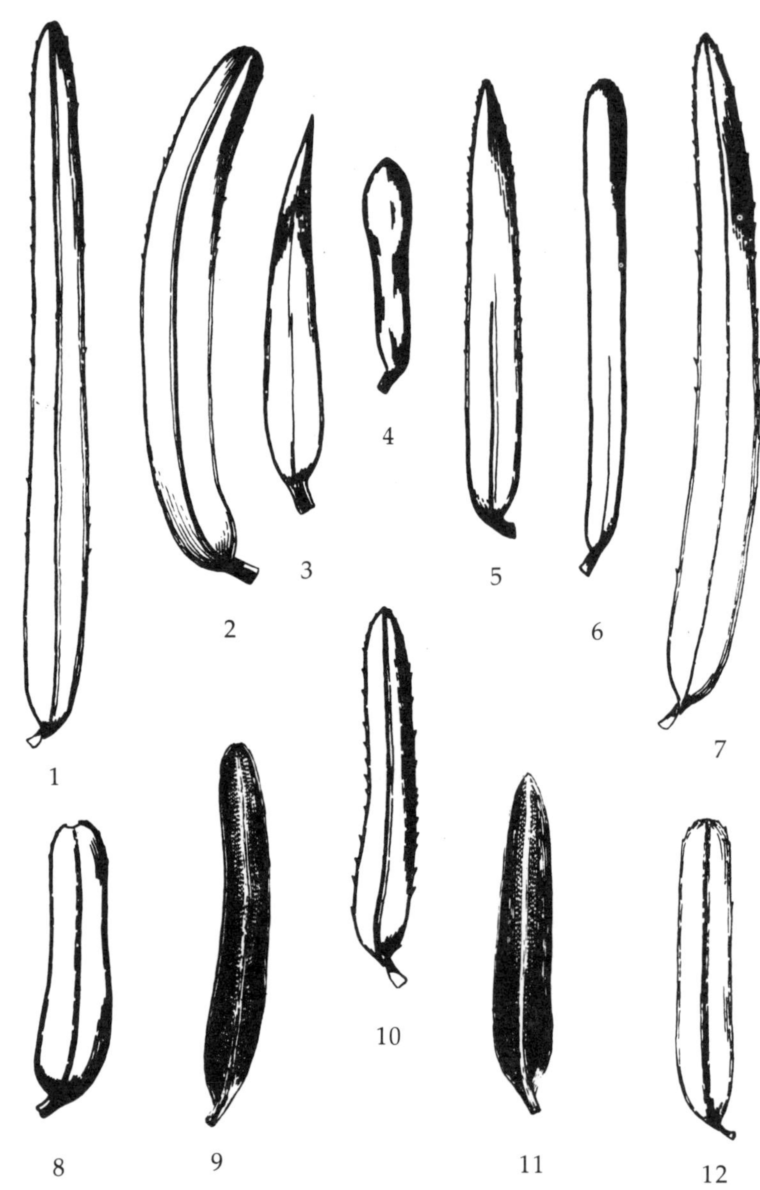

Variations in leaves of *Tsuga canadensis*. Magnification $\times$ 4 and one-half. In some cases the leaves are not typical of the specimens from which they were taken.

1. A long leaf, 21 mm in length, broadest below the middle and irregularly serrate along nearly all of the margin. 'Curtis Ideal'. J.C.S. No. 1317. 2. A curved leaf of nearly uniform breadth and serrulate only above the middle. 'Taxifolia'. J.C.S. No. 1369. 3. Broadest below the middle with a very acute apex and no serrulation. Similar to 'Cinnamomea'. J.C.S. No. 1433, unnamed variant lost to cultivation. 4. Broadest near the apex (nearly spatulate) and convex above, with no evident serrulations. Similar to 'Cinnamomea'. J.C.S. 1510, unnamed variant lost to cultivation. 5. Rather acute at the apex and finely serrulate. Similar to 'Curtis Sparseleaf'. J.C.S. No. 1375, unnamed variant. 6. A very narrow leaf, about ten times as long as broad and broadest above the middle. 'Angustifolia'. J.C.S. No. 1363. 7. Another leaf from 'Taxifolia'. 8. A very broad leaf, about four times as long as broad, and tending to be emarginate or notched at the apex. 'Wilton Wideleaf'. J.C.S. No. 1277. 9. An obtuse, nearly linear leaf with broad bands of stomata, five or six lines on each side of the midrib. 'Walton'. J.C.S. No. 1269. 10. Coarsely serrulate. J.C.S. No. 1496, an unnamed variant. 11. A leaf with narrow stomatic bands; three lines of stomata to the left of the midrib and four lines to the right. J.C.S. No. 1406, an unnamed variant. 12. A very blunt leaf, obtuse at the apex. Otherwise an apparently typical hemlock. J.C.S. No. 1406.

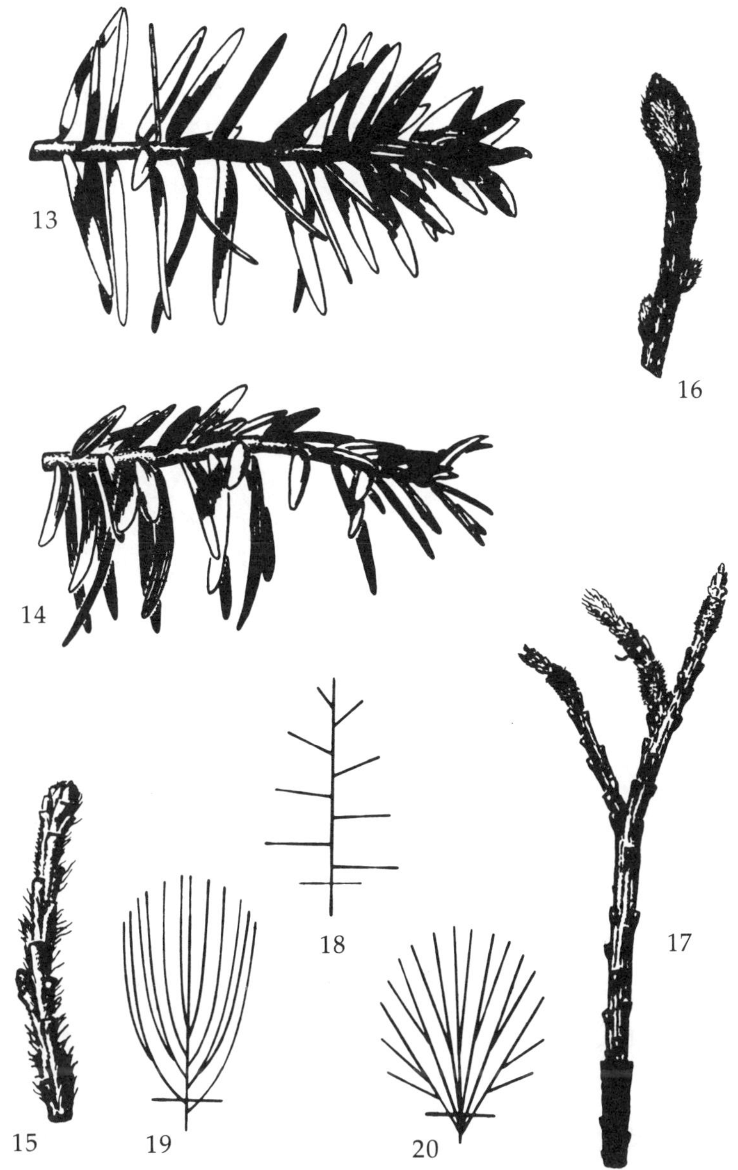

Growth characteristics of *Tsuga canadensis*.
13. Illustration of the radial disposition of leaves. 'Pumila'.
J.C.S. 1902. 14. Illustration of the two-ranked disposition
of leaves above but scattered below. 'Von Helms Dwarf'.
J.C.S. No. 1227. 15. A pubescent twig with a puberulant
(short pubescent) lateral bud on the left. This is typical of *T.
canadensis* and most variants of this species. Also from 'Von
Helms Dwarf'. J.C.S. No. 1227. 16. A nontypical twig of
Canada hemlock, appressed pubescent with pubescent
buds. A CINNAMON-TIP hemlock. J.C.S. No. 1368, an
unnamed variant observed in the Yonkers Nursery,
Yonkers, New York, in 1938. Lost to cultivation.
17. Showing the enlargements at the tips of twigs of
'Cinnamomea', well covered with cinnamon-brown hairs.
J.C.S. No. 1348. 18. A diagram of the excurrent habit of
typical hemlock. 19. A diagram of the fastigiate character.
The 'Schramm Fastigiate' deviates from this in that only the
lower branches are fastigiate.   20. A diagram of the
multiple-stem habit, as in 'Laurie' and 'Bagatelle Globe'.

Very interesting atypical oddities have appeared on the scene. One is 'Curly', introduced many years ago by Harold Epstein, Larchmont, New York, and another is 'Verkade Recurved', introduced more recently by Verkade's Nurseries, Wayne, New Jersey. The tips of the leaves of these odd plants curve downward and under like a claw. In 'Verkade Recurved' the fold is somewhat more appressed than in 'Curly'. The latter has an upright habit. The former is a bushy dwarf. Another odd cultivar is 'Kingston Hollow' with short and aborted leaves that are also twisted. This was collected by Walter Kolaga, proprietor of the one-time Mayfair Nurseries, Nichols, New York, in his own meadow while hunting.

Examples of the irregular disposition of leaves in hemlock variants. From left:
A Large-leaf specimen from a tree in Walton, New York, which is sometimes called the Walton hemlock. This tree is more than one-hundred years old. J.C.S. No. 1269.   A Dense-leaf variant. 'Von Helms Dwarf'. J.C.S. No. 1227.   A Conical to Pyramidal. Moderate Growth, variant: 'Curtis Ideal'. J.C.S. No. 1317. Also has long leaves, giving an effect of delicate foliage.   Treated in the Hemlock Thesis as the typical specimen of 'Sparsifolia'. This variant is no longer in cultivation.   A more or less weeping variant, with leaves quite radial on the contorted upper branchlets. J.C.S. No. 1273. This variant is not in cultivation.   A variant with leaves shorter than the typical hemlock. Treated in the Hemlock Thesis as var. *milfordensis,* which was in error. Originated in the nursery of George L. Ehrle, Clifton, New Jersey. J.C.S. No. 1362, no longer in cultivation.

COLOR

The most common color variation from the normal green in Canada hemlock is white. In most cases the spring shoots are green with the second or summer growth white. The white color fades through the winter and is green by the following spring.

Whitish hemlocks, especially WHITE-TIP variations, are very common. 'Dwarf White-tip' combines a dense compact habit when young with contrasting white tips in summer, 'Betty Rose' is one of the best of hemlocks, very dwarf, irregular, and picturesque, with short, clear white tips in spring and early summer. 'LaBar White-Tip' is an example of a faster-growing plant. 'Snowflake' has the longest whitish branchlets of any variant the writer has observed. On most WHITE-TIP hemlocks only the summer growth is white. This fades through the winter and is green by the following spring. Two variations do not conform. One is 'Frosty'. More than three-quarters of the otherwise normal foliage is whitish in summer and cream in winter. It forms a loose mound about as broad as high. The other is 'Betty Rose', just mentioned, which does not produce summer growth. This is the most dwarf of the WHITE-TIP hemlocks. 'Watnong Star' also shows white tips on the spring growth but the color is more fleeting.

A yellow-green color may be physiological, in which case, if transplanted to a different site, the leaves will assume the normal color. Even in normal cases, the color and depth of the green foliage is much affected by the soil conditions. The best known yellow or golden hemlock is 'Everitt Golden', which combines slow growth, dense leaves, and a bright golden color at certain times of the year, particularly in the spring and early summer. The cultivar with the slowest growth rate combined with yellowish foliage is 'Merritt Golden' in the collection of the late Henry Hohman, Kingsville,

Maryland. It is almost as slow growing as 'Minuta', which it closely resembles except for color. Mr. Hohman had no record of the source of this cultivar.

There are at least three cultivars which are extremely dark green. These are 'Boulevard' in the COMPACT GROUP, 'Westonigra', a fast growing, sparsely branched plant that has defied classification, and 'Armistice', a broad conical plant of slow growth.

STOMATIC REGIONS

A plant labelled *Parvifolia* was observed at Far Country in 1938 but has long since disappeared. Although the growing conditions were excellent the bands of stomata were dull. As with most plants with blue stomata the intensity of coloration is greatest in fertile soil. Other LITTLE-LEAF variants, such as 'Jenkinsii', have pale stomatic bands; therefore there is some correlation between short leaves and pale bands. A number of variants display intense bands of stomata. Many of these are related to the DENSE-LEAF and LARGE-LEAF GROUPS. There is some variation in the width of stomata. An example of narrow bands can be observed on the original plant of 'Dawsoniana' in the Hunnewell Arboretum, Wellesley, Massachusetts. Most leaves have three to four lines on each side of the midrib. Exceptionally broad bands can be observed on a large plant with tufted foliage in the Bayard Cutting Arboretum, Oakdale, Long Island, New York. Here each band consists of eight or nine rows of stomata. Narrow bands of stomata have frequently been observed in plants of the YEW-LIKE GROUP and in 'Brandley'. It is interesting to note that narrow bands usually consist of as many lines as the broader bands. Narrowness is usually associated with the crowding of the lines rather than by a reduction in their number.

The width and intensity of the stomatic bands, and in a few cases the number of lines of stomata that can be distinguished, vary from plant to plant, so this can in certain clones be a useful means of identification, but in most cases the number of lines (3 to 5) remain constant whatever the width of the band. The intensity of color varies from clone to clone but is much affected by growing conditions and environment, so it has but little diagnostic value.

GENETIC VARIATION

The extreme variability and the complexities of that variability in Canada hemlock have been previously emphasized, but we know little or nothing about the genetic nature of the individual variants. Only careful and extensive breeding experiments continued for many years would divulge the secrets and they would have little practical value unless superior, pure strains of forest trees could be isolated—an unlikely event.

Certain plausible explanations may, however, be offered on the basis of the behavior of other plants. Undoubtedly, the most important is the theory of mutation. Every cell of living tissue, whether in animal or plant, contains minute threadlike structures called genes. A complete understanding of how these function would admit us to the understanding of life itself, but it is clear that (probably acting in combination rather than individually) they control the whole organism throughout its life. They control its growth and its reaction to its environment, to damage and to disease. They govern it in maturity and in its efforts to reproduce itself and they determine the characteristics that it shall pass on to its offspring. To borrow a term from the jargon of engineers, they constitute its "blueprint."

Reproduction by sexual means is effected by the development, in the flowers of a plant or in the sexual organs of an animal, of separate male and female cells each containing only half the normal complement of genes. Fusion of two such cells at the moment of fertilisation produces a cell with the full number of genes, and this is the commencement of a new individual. In a plant, the development must pass through a phase (the ripened seed) which is of great value to the species, since the new life at that stage is exceptionally resistant to hazards from outside and is able to remain dormant, often for a long period, until environmental conditions are favorable for its germination and further growth. (And this life, "blueprint" and all, may be contained in a seed so microscopically small as to make man's efforts to develop and miniaturise his computers look very clumsy indeed.)

Each such new individual has its own "blueprint", compiled from the genes inherited from its parents. In a stable species this will be exactly as before, but should either or both parents have experienced a mutation that has altered its genetic make-up that change must in some way show up in each new "blueprint". These deviations control the new plant in accordance with natural laws which in the science known as Mendelism have been studied, and their action can be experimentally demonstrated in simple cases. But life never is simple: in a species regarded as variable these genetic peculiarities may have been accumulating over many generations. Thus the new "blueprint" will be so complicated as to render its analysis impossible and the effect it will produce unpredictable—sometimes surprising. The studies already referred to distinguish between the characters they designate dominant and those they term recessive. The former show up in the new plant; the latter do not, yet they are in its "blueprint" and must be transmitted to its offspring—often to appear in a later generation. It is the same phenomenon exactly as the familiar case of a child taking after a grandparent, an uncle or even a distant relative.

VARIATION FROM SEED

This apparently hap-hazard re-selection of genes differs in the "blue print" of each separate seedling, and this explains why certain forms come true from seed, others do not, whilst in many cases the crop of seedlings covers the whole range between complete normality and the full re-appearance of the peculiarity of the mother tree. There are many recorded cases of such variation from a single collection of seed. A few must here suffice.

Dr. A. B. Stout (*Journ. New York. Bot. Gar.* **40**: 162. 1939) collected seed in 1916 from a typical specimen about 4-1/2 feet high at the New York Botanical Garden. He tells us:

> About 25 seedlings were obtained. All but one were pendulous and six years later these varied in height from 10 inches to two feet. Some were flattened in form while others were more erect and rounded in profile. The only one which is not pendulous has a large main stem and one smaller one arising from the ground and both are erect and excurrent with a habit of growth that is normal. The largest of the others is now almost four feet tall and the smallest is three feet tall.

In 1924 Richard Guldemond of the Blue Ridge Nurseries, Harrisburg, Pennsylvania, collected seed from a compact hemlock in the Gettysburg National Cemetery, Gettysburg, Pennsylvania. He raised two or three hundred plants, and these were observed in 1938. All these plants were more or less slow growing and compact and not one could be called typical hemlock. The larger plants were slightly more than 3 feet high and smallest ones less than 1 foot high. The habit varied from globose to pyramidal, and the foliage from sparse to dense, and from two-ranked to radial, in arrangement.

In 1926 Valleau Curtis of Curtis Nurseries, Callicoon, New York, sowed a considerable amount of hemlock seed in outdoor beds. The vast majority of the plants grown from this seed were typical Canada hemlocks. A goodly number, however, perhaps a fraction of 1 per cent, were definitely different from normal hemlock in one or more characters. There were many kinds of variations representing a considerable difference

Two 'Sargentii' seedlings in the South Wilton Nurseries, Wilton, Connecticut.

in growth rates, habit, and foliage characters. This one batch of seedlings produced more useful variants than were observed in any other nursery. A number of them have been selected and given provisional names. (See p. 100 under 'Curtis').

M. G. Coplen of the Rock Creek Nurseries, Rockville, Maryland, contributed another interesting example of diversity. In 1921, while travelling slowly along one of the narrow country roads near Mt. Storm, Grant County, West Virginia, he noticed two hemlocks in an isolated dooryard. The larger tree, almost 40 feet high, was apparently a typical Canada hemlock. He investigated and found a number of odd-looking seedlings outside the fence. He located the owner and obtained permission to dig the small seedlings, about twenty in number. Of the twenty about twelve survived. The foliage characters are much like typical hemlock but there are great differences in habit and rate of growth. The height ranged from 1 to 12 feet. The two hemlocks were isolated from other stands of hemlock, therefore the seedlings collected by Mr. Coplen must have originated from one or both of these two plants. The possibility of crossing must be admitted. Grafted plants of considerable size of these different variations were given to the U.S. National Arboretum by Mr. Coplen before his death.

More recently, in 1953, seed was gathered from coning plants at the U.S. Horticultural Field Station at Beltsville, Maryland, and germinated at the U.S. Plant Introduction Station at Glenn Dale, Maryland. When three years old, they were set in nursery rows 4 feet apart and 2½ feet apart in the rows. At the age of six years there were one hundred and thirty-four plants ranging in size from 16 to 20 inches in height and 36 to 45 inches in breadth except for one very dwarf plant 5 inches high and 18 to 26 inches wide (Ackerman and Seaton, *Amer. Hort. Mag.* **43**(3): 181. 1964.)

These cases are typical of the variation that can occur from seed collected from a single mother plant, but equally as many other instances can be cited to indicate that some mutation is, as it were, beginning to settle down genetically.

It might be expected that the Sargent hemlocks would come more or less true from seed since four or five similar plants were originally found, presumably in one spot and all having had the same mother, but there are many other examples of like begetting like under natural conditions with no help from the hand of man. One striking example is the diminutive clone 'Minuta'. Daniel M. St. George of Charlotte, Vermont, discovered a number of plants of this very dwarf variant in 1927 in an undisclosed location in the Green Mountains. Mr. St. George visited this locality several times and altogether transplanted about twenty-five of these tiny dwarfs to his nursery. He observed a larger plant near the seedlings, 2 feet high and bearing cones. Another extreme dwarf, 'Abbott's Pygmy', was discovered as a group of three apparently identical seedlings by Frank L. Abbott north of Richmond, Chittenden County, Vermont.

The late Wm. C. Horsford found a tree 50-60 feet high in a pasture in Vermont which had twisted branches, contorted branchlets and 'Cinnamomea'-type foliage and which was surrounded by seedlings with similar characteristics, one of which was selected for propagation of the 'Horsford Contorted' clone.

'Youngcone' provides another similar example. According to Llayne Ziegenfuss Hillside Gardens, Lehighton, Pennsylvania, there are numerous natural seedlings in a pasture. Many of them have multiple stems and are upright to globose in habit, mostly compact. Most of them produce cones while very young, hence the name. Scions were taken from one of the better seedlings.

A point of some difficulty arises if named cultivars are reproduced from seed. Although the *International Code of Nomenclature for Cultivated Plants* recognises a cultivar "consisting of one clone or several closely similar clones" the International Registration Authority is strongly in favour, in the case of such long-lived plants such as conifers, of limiting the use of the cultivar concept so far as possible to a single clone. The term clone is defined as a "genetically uniform assemblage of individuals derived originally from a single individual by asexual propagation." However similar the seedling appears to be to its mother, even if it is indistinguishable, it is beyond question not the same clone. Seedlings should therefore on no account be distributed under the cultivar name of the parent tree.

MUTATIONS:

WITCHES' BROOMS

The occurrence of mutations is probably extremely rare—the proverbial one in a million chance. Any such that are genetic in origin cannot be detected until they show up in a later generation. Mutations occurring in a growing plant may be so slight as to pass unnoticed, but they are forced upon our attention when they cause the conglomeration of congested growth which we call a witches' broom. These peculiar growths (German: *Hexenbesen*; French: *Balais de sorcière*) are due to the abnormal development of buds. It is certain that every branch of a woody plant has many more bud initials than will ever develop. These are held in reserve, and if the terminal growth is cut, any of these auxiliary buds may be stimulated to produce new growth. But occasionally many of these buds on a particular branch of a particular tree develop without any discernible external influence. In one way or another, the cambial cells which are responsible for initiating the development of buds are given extra stimulation. The result is abnormal, congested growth. These adventitious buds may be produced near the nodes or in the internodes. When all of these buds develop, the growth is very dense.

In addition to being densely branched, the growth of a witches' broom is usually upright, which of course helps to produce the broomlike effect. We can understand such conditions if a parasitic organism is present, for the organism interferes with the normal development of the plant. The witches' broom is a response to the presence of the organism. If the invasion is general, the life of the plant is sometimes shortened.

These occurrences have long caused interest and have been studied by many observers, at home and abroad. Welch (*Manual of Dwarf Conifers* 12. 1979) has given a resumé of current knowledge and he includes a good bibliography which need not be repeated here. In many groups of plants broom-growth can be traced to particular parasites, fungi or some other pathological origin, but in the case of conifers no such causal mechanism can be found.

Vegetative propagation of shoots from one of these brooms will frequently develop into attractive, slow-growing plants with a charm of their own and such form attractive garden plants. They are usually stable, but the possibility of further mutation occurring is always present and must be watched for vigilantly, especially if it takes the form of a reversal of the original mutation. The resulting vigorous growth will soon take over and destroy the attractive character of the plant, if not cut away promptly.

Quite a few of these brooms have been observed by the writer on Canadian hemlocks.

Most brooms occur on lateral branches instead of the terminal. A most striking terminal broom was discovered by the author in 1938 along the Elmira Road about one-quarter mile west of the Lower entrance to Enfield State Park, New York. It formed the whole top of a 35-foot-high plant and was about 6 feet high and 6 feet wide with compact growth due to abnormal bud development. The foliage was close to typical hemlock.

Another terminal broom was observed in 1938 on the estate of Robert H. Montgomery, Cos Cob, Connecticut. This was also growing on a small hemlock about 4 feet high, but the habit was nearly globose with a small nest-like concavity on one side of the top. It was rather compact with leaves shorter than normal and foliage of a sickly green, perhaps caused by some organism.

A number of witches' brooms have come to light since 1939. Fred Bergman has propagated and named at least four: 'Fantana Broom', from a sizeable broom on a plant of 'Fantana' that had been sold to a friend: 'Helene Bergman' from an atypical broom on a plant of 'Cole', very dwarf and flat: 'Bonnie Bergman', from a 2½ foot tree with a spreading terminal broom; and 'Barry Bergman', with foliage like 'Bennett' (but in habit it forms a plant nearly as high as broad) from another broom-like structure on a plant of 'Cole'. The original plant of 'Bonnie Bergman' came from a Christmas tree grower near Stroudsburg, Pennsylvania, in the fall of 1973.

In the Arnold Arboretum collection there were small plants of an unusual graft from Joe Cesarini, Ridgely, Maryland, resembling 'Cole', from a broom near Hicksville, Long Island, and of 'Narragansett' a witches' broom discovered by James W. Brandley near Providence, Rhode Island. In various collections are plants of 'Kathryn Verkade' from Verkade's Nurseries, Wayne, New Jersey. Several unnamed, untested brooms from various locations have been propagated by Llayne Ziegenfuss, Lehighton, Pennsylvania.

There is only one report of a witches' broom being found on *T. caroliniana*. Gordon E. Jones, Director of Planting Fields, Oyster Bay, Long Island, New York, found three different brooms between Exits 54 and 55 on the Connecticut Turnpike. He sent scions to Arnold Arboretum in 1965, but there is no record of their survival. This disparity between the frequency of broom development in the two species ties up with the correlation there evidently exists between the propensity to develop witches' brooms and the much greater variability that characteristises *Tsuga canadensis*.

A witches' broom at South Wilton Nurseries, Wilton, Connecticut, in 1938. No longer in cultivation.

T. canadensis 'Bonnie Bergman', a witches' broom. Raraflora, Feasterville, Pennsylvania, 1944.

Pinus banksiana seedlings from cone-bearing witches' brooms raised at Arnold Arboretum, Jamaica Plain, Massachusetts.

Pinus strobus witches' broom seedlings. Note the wide variation in sizes.

GENETIC DWARFISM

Bud mutations have a profound effect on the character of any species, especially with regard to its variability, so much so that we cannot avoid the conclusion that there is some correlation between witches' broom formation and a multiplicity of variants, such as exists in the case of *Tsuga canadensis.* It is unusual for a witches' broom to develop cones, but when one does the result is, as wou'd be expected, a progeny of seedlings widely varying in the extent to which the dwarfism of the mother has been inherited. Numerous cases are on record of such an occurrence in several species of *Pinus, Abies* and *Pseudotsuga* but I personally only know of one instance of seedling propagation from a witches' broom on hemlock. In December 1973 the author visited a nursery, Phytoecology, in Ridgely, Maryland. There they had grown thirty-two seedlings in gallon cans, about fifteen years old at the time, from cones on a broom found with viable seeds near Hancock, New York.

Twenty-seven plants were observed, the smallest the size of a baseball, the largest about 8 inches high and 6 inches wide. The intermediate sizes resembled 'Minuta'. A couple were more regular than 'Minuta', presenting a clipped appearance.

More important, probably, is the effect of apparently normal trees secretly carrying recessive genes—and therefore being able to produce out-of-character progeny. This matter of genes dormant in a plant but capable of asserting their effect in later generations has been already referred to, and it is interesting to speculate that many of the tiny plants that appear in forest nurseries, in ornamental nurseries, and in the open forest or meadow may inherit their dwarf character from some witches' broom that flourished and bore seed a thousand years ago. Of course this need not always occur. Mathematically, each successive generation that avoids it reduces the likelihood that it ever will show up; after innumerable generations the possibility will be infinitesimally small and a new species will have appeared. But it would seem that a good measure of stability can be very soon established, because many cases are recorded of the discovery of a group of unusual but all very similar trees—a "population" so local as to indicate a single mother tree. Also, many known variants come true, or nearly so, from seed, the well-known case of Sargent weeping hemlock being one of these.

An early account of experiences in growing Sargent hemlock from seed was contributed by A. B. Stout (*Journ. New York Bot. Gar.* **40**(15): 162. 1939). He quotes from a letter he had received from Jacob C. van Heiningen, proprietor of South Wilton Nurseries, Wilton, Connecticut. Mr. van Heiningen wrote: "I beg to inform you that the 2,000 hemlocks we raised from seed were from seeds collected in 1906 from two Sargent weeping hemlocks which were growing near the office of the Parsons & Sons Nursery Company, Flushing, Long Island, at that time but since cut down by real estate operators. I shipped the seeds to my brother in Holland and they grew 100 per cent weeping. After they grew to a size of one to two feet they were sold in large quantities in Europe and the U.S.A. In the spring of 1918, the last year we could import to the U.S.A. I received 150 plants. It is curious that there is so little trace of all these plants in Europe today.

Seedlings of the cultivar 'Brookline' at Arnold Arboretum. Note the uniformity of habit, which is usually prostrate in this cultivar.

5 CULTIVARS OF CANADIAN HEMLOCK

EARLY ATTEMPTS AT NOMENCLATURE

This species had had a struggle to find its true generic identity, although the specific association with Canada has been recognised from an early date. First listed as *Pinus Canadensis* by Linnaeus, (*Species Plantarum*. ed. 3, **2**: 1421. 1763), it was transferred to his new genus *Abies* by Michaux (*Flora Boreali Americana* ed.3, **2**: 206. 1803) and finally to *Tsuga* by Carrière (*Traité générale des Conifères* 185. 1855).

The variability within this species was soon noticed, the first authority to mention variations in Canada hemlock being Francois André Michaux (loc. cit.). He stated that the hemlock spruce sometimes shows a peculiarity which he had not noticed in any other North American tree, namely the habit of occasionally ceasing to grow at the height of 2-3 feet. He referred to a dwarf plant in a certain locality near York Court House between Portland and Portsmouth, Maine, where the soil was dry and stony. He described this plant as thick and dense, inclined to be pyramidal with branches resting on the ground.

This started a long history of names being given to selected variants, the generic name *Abies* gradually giving way to the present name *Tsuga*. Carrière (*Man. Plantes* **4**: 334. 1854) made a direct reference to the plant mentioned by Michaux. He introduced for it the name *Abies canadensis nana*, indicating that dwarf seedlings must have been selected in Europe prior to 1854. The following year, in his more specific work on conifers, (*Traité générale des Conifères*, 190. 1855) he classified all hemlocks under the genus *Tsuga* and the above then became *Tsuga canadensis nana*, although Gordon (*Pinetum*, 15. 1858) kept to the old name, giving Knight as authority and Carrière's name as a synonym. Later (*Pinetum Suppl.* 9. 1962) Gordon describes *Abies canadensis gracilis* Waterer, which he records as having originated on a nursery in Surrey, England. Lindley (Gar. Chron. **20**: 460. 1864) added another cultivar, *microphylla*, described from a single plant.

In 1866 Nelson, in a more-or-less popular work entitled *Pinaceae: Being a Handbook of the Fir and Pine,* described *Abies canadensis* and referred also to *Abies brunoniana, heterophylla, mertensiana* and *sieboldii* as "quasi-species" stating that he found the distinctions vague. Of *T. canadensis* he remarked, (giving very brief and inadequate descriptions) that the only cultivars worth noting are *argentea, aurea, graciles, microphylla* and *nana*. Of these *argentea* and *aurea* appear for the first time.

In his second edition Carrière adds (249. 1867) a description of *gracilis* taken in from Gordon, but in the following year Sénéclauze, a French nurseryman of Bourg Argental, Loires, published his manual, *Les Conifères,* in which he introduces the names *compacta* and *latifolia,* both described from seedlings on his nursery. The former name has been widely and loosely used in many connections, but the latter has not been heard of since. Both clones are lost to cultivation. Sénéclauze also used the already occupied name *microphylla* for another of his seedlings which also appears to have been lost.

Gordon, in his second edition, gets a little confused about the epithet *microphylla* (which he cites as a synoym of *nana*) but he introduces *alba spica* Barron and *milfordensis* Young.

Josiah Hoopes, a prominent nurseryman near West Chester, Pennsylvania, published the first American book dealing exclusively with conifers in 1868, entitled *The*

Book of Evergreens, a Practical Treatise on the Coniferae or Cone-Bearing Plants. He listed two varieties of Canada hemlock (under *Abies canadensis),* var. *nana* Lawson and var. *microphylla* Lindley, and gave a full account of Lindley's original description and Michaux's experiences with hemlock in America. This author is also remarkable for having given us (p. 418) the first recorded description of the Sargent Weeping Hemlock, although without naming it. (See Chapter Five p. 134 for a full account). The name was not validly published until *The Art of Beautifying Gardens* (Scott, F. J.) appeared in 1870.

Otto (*Hamburg Garten und Blumenzeitung* **29**: 202. 1873) recorded the occurrence of color variations and applied to them the name *variegata,* an already occupied name.

After the Centennial Exposition of 1876 in Philadelphia, Pennsylvania, Joseph T. Rothrock published in 1880 a *Catalogue of Trees and Shrubs adjacent to Horticultural Hall.* The following varieties were listed under *Abies canadensis* (p. 15):

> var. *microphylla* Lindley, a hardy dwarf, leaves small, dark green and round on the edges; var. *milfordensis* Young, dwarf, globular in form, shoots slender, drooping leaves smaller than the species; var. *nana* Lawson, 2-3′, foliage spreading, tufty. Also var. *nana nigra* Hoopes: var. *hemisphaerica* Hoopes; var. *inverta* Hoopes; var. *pendula:* var. *Sargentii;* var. *Shotwelliana* Hoopes; var. *variegata* Hoopes.

It is evident from this that since the publication of his *Book of Evergreens,* Hoopes had introduced hemlock varieties from Europe and also had made selections from his own nursery seedlings. No further reference has been found to his varieties *hemisphaerica, inverta, nana nigra, Shotwelliana* and *variegata;* these plants have probably long since passed out of existence. Since his description of *milfordensis* repeats that given by Gordon, probably the plant in the exhibition was a very small one, too small to describe. It is the first time vars. *Sargentii* and *pendula* were distinguished.

Ludwig Beissner was one of the greatest students of conifers. In 1884 in co-operation with Jäger he published *Die Ziergehölze.* In this interesting volume were gathered up all the previously published names, all now correctly placed under *Tsuga. Pendula* appears in a European publication for the first time and with a description that clearly relates it to what I here regard as the European form of weeping Canada hemlock.

In 1887 an important Congress of conifer experts and growers was held in Dresden, Germany. Prior to the Congress, Beissner published *Systematische Eintheilung der Coniferen,* a check list of all conifer species and forms surviving in Germany in the open or under shelter, with their synonyms. In October of the same year, after the Congress, an amended check list entitled *Handbuch der Coniferen-Benennung* was published, in which the name *globosa* only is added, but in his major work *Handbuch der Nadelgehölze* (402-3, 1891) Beissner introduces, with descriptions, the epithets *compacta nana* Hort., *fastigiata* Hort., *columnaris* Bolle, *macrophylla* Hort (misidentified under *T. mertensiana)* and *fol. argent. varieg.* Hort., which appears to be indistinguishable from *albo-spica.*

In 1889 appeared the celebrated catalogue of the nursery firm S. B. Parsons and Sons, Flushing, Long Island, New York, which contains the description of their Sargent hemlock stock and how they came by it. Also listed is *Abies* (now of course *Tsuga) canadensis atrovirens.* A tree received in the Arnold Arboretum in 1880 under this name that is still growing there does not answer to the nursery description, but in view of its seniority it must be accepted as the clonotype. A living plant must have greater authenticity than any description taken from it!

In 1897 Sudworth, one of our best known American dendrologists, published the results of his study on the nomenclature of the trees of North America. Without any apparent justification he suggested new names for several of the already established clones (such as *argentifolia* for *argentea, compacta minima* for *compacta nana, erecta* for *fastigiata, paucifolia* var. *sparsifolia* and *pumila* for *nana.* Later authors have not recognized Sudworth's revision, and his names can only be charitably considered as quite unnecessary alternative spellings.

After this, the flow of new names slowed down. *Pendula argentea* appeared in the *Kew Hand-list of Conifers* (ed.2, 71. 1903) and here for the first time var. *pendula* and var. *Sargentii pendula* are clearly separated. In 1900 Kent in Veitch, *Manual of Coniferae* (ed. 2, 465. 1900) had used the name *Sargentiana* apparently with no particular plant in mind, since he thought it to be the *nana* of European gardens.

The German writer Schneider in Silva-Tarouca, *Unsere Freiland-Nadelgehölzer* was the first writer to introduce any system of classification. He (303. 1913) used the two infra-specific botanical categories, *varietas* and *forma*, and established var. *pyramidalis* to include "forms of upright habit" and var. *nana* to include "forms of dwarf habit". Using that system he described several varieties as follows:

1. Var. *albospica*	Color forms
2. Var. *nana*	f. *compacta*
	f. *globosa*
	f. *pumila*
3. Var. *pendula*	f. *Sargentii*
	f. *nana pendula*
4. Var. *pyramidalis*	f. *columnaris*
	f. *fastigiata*
5. Short-leaved	f. *gracilis*
	f. *microphylla*
	f. *paucifolia*

Bean (*Trees and Shrubs* 606. 1914) maintained the distinction between the hemi-spherical var. *pendula* and the American var. *Sargentii*. He mentioned two other variants and referred to the existence of others. In ed. 3, (Vol. III. 441. 1921) he added 'Prostrata', describing it from a plant in Scotland that is still there. (But see p. 66.)

In 1923 Hornibrook's *Dwarf and Slow-Growing Conifers* was published and provided the first widely available account of the finding and distribution of the four original Sargent hemlock seedlings. Unfortunately he makes the statement often quoted since about two of these trees being dead (of which there is, so far as I am aware, no corrob-oration) and assumes that a statement by Professor Sargent, i.e. that all the nursery-men's stock (i.e. in America) were Sargent hemlocks applied also to the European form, which he then carefully describes as "the best form". He also mentions 'Wellesliana' as a small bush 4 feet high but without naming it until his second edition.

In 1930 Fitschen revised Beissner's *Die Nadelhölzkunde* and made another attempt to classify the varieties within the different species of cultivated conifers. For *T. canadensis* he established three main classes as follows:

I. Variations in growth
   1. Low forms: var. *nana* Carr., var. *compacta nana* Hort., var. *globosa* Hort., var. *pumila* Ordnung and var. *atrovirens* Hort.
   2. Columnar forms: var. *fastigiata* Hort. and var. *columnaris* Bolle.
   3. Weeping form: var. *pendula* Hort.
II. Variations in leaves
   var. *sparsifolia* Hort., var. *gracilis* Carr., var. *macrophylla* Hort., var. *microphylla* Hort. and var. *parvifolia* P. Smith.
III. Variations in color
   var. *aurea* Hort., var. *argentea* Hesse, var. *albo-spicata* Hort., and var. *argentea-variegata*

Chittenden (*Conifers in Cultivation* 546. 1932) introduced var. *brevifolia*, a plant growing in Ireland, but no description was included. Baily (*Cultivated Conifers* 124. 1933) included twelve varieties previously noted and described four more varieties for the first time: *Dawsoniana* Bailey, *Fremdii* Koster, *Hussii* Hort., and *Jenkinsii* Bailey.

Bean (1933, p. 461) added another name, var. *prostrata,* based on a plant in the garden of Mr. Renton, Branklyn, Perth, Scotland. He clearly distinguishes between the European form (var. *Pendula)* which he accurately describes, and var. *Sargentii* which he says "forms a compact pendulous-branched shrub 6 feet or more high". A Sargent hemlock this high would be quite an established specimen: the plant seen by Bean was a much younger plant, planted on a steep slope where its pendulous tendency would give it an appearance of having a prostrate habit, but its subsequent development has made it clear that it is a Sargent hemlock, possibly the clone 'Brookline'. The name is therefore redundant, and should no longer be used.

Teuscher (*New Flora and Silva* **7:** 274. 1935) was responsible for naming a very dwarf plant as var. *minuta,* based on one of a group of seedlings discovered in the Green Mountains near Charlotte, Vermont.

Den Ouden included var. *wellesliana* in his list of hemlock varieties in the Dutch work *Naamlijst van Coniferen* (46. 1937) and the following year, in his second edition, Hornibrook described it as low conical bush of pendulous branches, about 5 feet high by as much through, a seedling "at the Wellesley Pinetum, Massachusetts". If by this he meant the Hunnewell Arboretum, Wellesley, Massachusetts, the tree can no longer be found.

EARLY COLLECTORS

By this time an interest had developed in the garden value of the mutations in this species, and a small band of enthusiasts were collecting diminutive and other forms from the wild. Usually these were recorded under numbers, many of which have since received cultivar names.

Frank L. Abbott was a lifelong hunter, fisherman and woodsman. He lived at Worcester, Massachusetts but had a summer cottage in Athens, Vermont. He was interested in the mutations of the Canadian hemlocks and was responsible for finding quite a number variants. He identified his finds by numbers. A few of these have received cultivar names; others may be lost to cultivation. They were never adequately described, but if any of the unidentified numbers survive in collections (and are regarded as sufficiently distinctive) the author would be pleased to hear about them.

In the 1920s Ralph Bacon, a retired landscape gardener of Sparta, New Jersey, found a number of very dwarf, dense and irregular seedlings in the wild. He transplanted nine of these, giving each a number. Of these, five of the survivors are still to be found in collections, but so far as I am aware only one of the Bacon series (Mk 1—'Bacon Cristate') has ever been registered.

From a large batch of seedlings raised in 1926 on the Curtis Nursery, Callicoon, New York, a number of interesting slow-growing selections were made, several of which are in cultivation. Similarly, Joseph F. Gable, a nurseryman of Stewartstown, Pennsylvania, produced a numbered series of several selected forms. These, along with some of his own introductions (identified as the Kingsville series) were propagated and distributed by the late Henry J. Hohman, proprietor of the Kingsville Nursery, Kingsville, Maryland.

But from all these numbered series we inherit two problems. Firstly, none of these introducers issued catalogues, (or if they did they were undated and without descriptions). Secondly, an unfortunate habit developed of replacing the Series numbers, as each clone became known and needed a name, with a two-word name consisting of the originator's name plus a classificatory word such as 'Little-leaf' or 'Spreader'. This potent source of confusion is dealt with on p. 84. We are stuck with such names as have been validly published, but any extension of the practice is to be deprecated. Where fresh names need to be found, names in a form more acceptable for registration should be put forward.

Later, a younger generation of nurserymen and plant collectors arose. Of these, Layne Ziegenfuss of Hillside Nursery, Lehighton, Pennsylvania, together with his friend Gregory Williams went more or less thoroughly over the territory covered by the writer during his tour of 1945 and identified and brought into production many of those early selections that might otherwise have been lost. The late Fred Bergman of Raraflora Nursery, Feasterville, Pennsylvania distributed a wide variety, including several of his own introduction, and there were several others. These were mostly in the Eastern States, but a number of wholesale nurseries on the West coast are now building up their collections and stocks in anticipation of growth in the demand for the garden-forms, a trend that this book should help to develop.

Not only have there always been avid collectors and specialist nurserymen, a very large number of the cultivars we have today were originally the discovery of someone not previously interested in hemlocks. In many cases a single cultivar is a person's sole contribution. Listing these contributors here is not feasible, but the name of the originator (where known) is given for each cultivar in the descriptions later in this chapter.

CHARLES FRANCIS JENKINS

Charles Francis Jenkins was born 17th December 1865 at Norriston, Pennsylvania. His early boyhood was spent at Wilmington, Delaware, but he obtained his schooling at West Chester, Pennsylvania. When 18 he entered the employment of his maternal uncle, Wilmer Atkinson, founder of *The Farm Journal* in Philadelphia. He retained his association with this paper all his life, becoming Chairman of the Board in 1935.

In 1890 he married and of four children, one son and a daughter are still (1983) alive. He was interested in the history of Pennsylvania and was a Member (and in many cases President or Secretary) of a number of societies, both local and national. He was also involved in considerable commitments with charitable and educational establishments, including Swarthmore College, Swarthmore, Pennsylvania, where he was elected to the Board of Management in 1904, became Vice-President in 1920 and Chairman in 1933. He died 2 July 1951 at the Pennsylvania Hospital, after a long illness.

Arising doubtless out of his lifelong business involvement with agricultural publishing he became a keen horticulturist. At his home "Far Country", Germantown, Philadelphia, Pennsylvania he established the Hemlock Arboretum in 1931 and there he developed an unique collection of species and cultivars of all the cultivated hemlocks he could lay hands on, as well as many other rare and notable plants. He also edited and distributed the *Hemlock Arboretum Bulletin* from 1931 until his death, Bulletin No. 74 being the last of the series to be issued.

THE HEMLOCK THESIS OF 1939

Having gathered thus far the attempts at naming the garden forms scattered throughout the literature of both America and the Old World, the time has come for me to introduce a remarkable figure, Charles F. Jenkins, who had great influence in promoting interest in hemlock variations. A prosperous business man, at the age of sixty he was afraid of going stale so he sought for an absorbing interest for his later years and found it in the study of hemlocks. In the grounds of his residence "Far Country", Kitchens Lane, Mt. Airy, Germantown, Philadelphia, he established the Hemlock Arboretum. From purchases and gifts from nurserymen and collectors in this country and in Europe he gathered by far the largest collection of *Tsuga* species and variants that had ever been attempted. He also carried on a wide correspondence and encouraged visitors, most of the principal figures of the time in American horticulture and many visitors from overseas having found a ready welcome there.

He also edited and distributed each quarter from 1932 until his death in 1951 *The Hemlock Arboretum Bulletin* in which he left a record of his many hemlock acquisitions and published many interesting items of Hemlock lore. These *Bulletins* will be frequently cited throughout the present book. They are by no means a complete record, but as a history of the growth of interest in hemlocks they are both interesting and informative.

When this unique collection of hemlock variations was established at Far Country, the wealth of material was demonstrated and its value was soon officially recognised. In the *Agricultural Yearbook* of the U.S. Department of Agriculture for 1949, *Trees,* an article (p. 449) designed to encourage private collections of trees singled out Mr. Jenkins' Hemlock Arboretum as one good example. By 1937 there was such an assortment of plants in his collection that Mr. Jenkins sought technical help. It had become evident to him that a project to assemble, classify and evaluate all the available variations of hemlock would be of much value, and in the *Hemlock Arboretum Bulletin* 19 (1937) he wrote:

> Is there not some young man (or woman) who has graduated from college in botany, horticulture, biology, ecology, dendrology, landscape gardening or some kindred science who would like to work on a thesis for a doctor's degree on the "*Mutations of Tsuga canadensis*"? It would be a useful, worthwhile subject, a real contribution to scientific knowledge, and fortunately there is now plenty of laboratory material.

Bulletin No. 22 (1938) carried this statement:

> In *Bulletin* No. 19 it was stated there was an opportunity for some young man (or woman) who might want to work for a higher scholastic degree to use the Arboretum and the records which have so far accumulated as a basis for the necessary thesis. I am happy to announce that John C. Swartley has undertaken the job with the advice and consent of his preceptors of the Department of Floriculture

and Ornamental Horticulture of Cornell University where Mr. Swartley is now enrolled for advanced study. He will confine his research to the mutations of *T. canadensis,* feeling that this subject will be big enough and broad enough to make a worthwhile contribution to botanical and horticultural science.

Mr. Swartley graduated from the University of Pennsylvania and since 1933 has been doing practical work, propagation, plant and seed collecting and plant identification at the Morris Arboretum, Chestnut Hill. His training and enthusiasm will equip him for his undertaking. I ask for him the co-operation of all hemlock lovers. He has already made a good start in classifying the hundred or more specimens at Far Country, bringing the records and measurements up to date and checking nomenclature and labels. He will undoubtedly make discoveries of new variations.

The way was thus opened for the author to take up an interest that was to remain with him for life. It would result in a Thesis with the title *Canadensis Hemlock and its Variations* being submitted to Cornell University in September 1939 and in a supplement prepared in 1945. Neither the Thesis nor the Supplement were ever published. Many collectors and botanists have borrowed the Hemlock Thesis either from the writer or from Cornell University Library. Several authorities have lauded the Hemlock Thesis as the only definitive work on variations of Canada Hemlock.

TECHNIQUE USED

The title given to me had in the first place to be interpreted as meaning variations in each part of the tree, taken separately. I therefore embarked on a massive compilation of statistical data based on measurement and counts of such matters as leaf length, leaf width, density (i.e. number of leaves per cm. of shoot) and similar factors. Interesting techniques were worked out to reduce the time taken by this chore to the minimum. A system was devised to make definite comparisons between certain characters of hemlock, namely length of leaves in millimeters and density of leaves on the branchlets expressed in number per centimeter. Slots with lengths varying from 3 to 19 milli- meters were cut in the top of a box and so arranged that a single match-box could be secured beneath each slot. Pieces of growth from one and two-year wood on the lateral branchlets were then selected and photographed. After the leaves dried sufficiently to detach themselves from the branchlets, the operator dropped them singly through the proper slot, using forceps. In this way, the frequencies at intervals of one millimeter were directly obtained, and the mean or average length of leaf easily calculated. After adding the frequencies to obtain the total number of leaves and measuring the total length of branchlets, the number of leaves per centimeter was easily determined. Statistical methods were then applied.

Background information was obtained by searching through "the literature" especially the *Hemlock Arboretum Bulletins* for references to the Canadian hemlock, and from Mr. Jenkins' notes and from enquiries both in this country and abroad. Until about 1939, as noted above, most of the existing varieties of hemlocks had been named in Europe. Very little information was obtained from there even though numerous European nurserymen were approached through correspondence.

The primary need was for travel and field-work. During the spring and summer of 1938 thousands of miles were covered, travelling through New York, New England, and then south to Washington, D.C. Seventy-three nurseries, arborists, and private gardens and estates were visited. Many of these were revisited in 1945 and again in the early 1970's.

Hundreds of plants were recorded and studied, out of the many thousands seen during the period. Each tree selected for study was allotted a reference number, the trees where possible being permanently identified by a correspondingly numbered label. Many of these early collections were photographed, but attempts to make herbarium specimens were not successful, because on drying the leaves fall away from the shoots. Since 1970 specimens treated in accordance with the recommendations of Santamour and Kettlewood (1963) have been more successful. Anyone embarking on a program of making herbarium specimens of this genus (or of the spruces, *Picea*—equal offenders in the matter of leaf dropping), should read the whole paper but, briefly, the method is to submerge the whole specimen in a 50:50 mixture of commercial glycerine and water for three weeks, followed by a rinsing under the tap and several hours' drying in room temperature before mounting. Samples were taken before or after (i.e. not during) the growing season. Total retention of the leaves is recorded in the Paper as being well over 90%, being slightly better in the samples taken in Spring than the Autumn cuttings.

It was also clear that along with the variations in leaf form, etc., variations in habit and vigour between one plant and another equally needed to be studied. It is of course these that give rise to fastigiate and pendulous trees, variegations and dwarf forms, all of particular interest to horticulturists. In 1939 there were recognised rules of botanical nomenclature, and in the absence of any better guide it was usual to list these horticultural variants as *"varietates"* (abbreviated to "var."), pushing them willy nilly into the botanical system although in most cases they were indisputably horticultural selections, recognised nowadays as "cultivars". This term, an abbreviation of the words "cultivated variety" was the invention of Liberty Hyde Bailey (*The Cultivated Conifers*. 13. 1933). Professor Alfred Rehder recognised the same distinction by his use of the phrase "the clonal garden forms" (*Bibliography*. xi. 1949), but a clear recognition of the distinct needs of horticultural nomenclature embodied in a separate *Code for the Nomenclature of Cultivated Plants* only came years later.

The author had to conform to the nomenclatural limitations of the time and classify these variants as botanical *varietates*, but it is interesting, in the light of present-day practice of regarding them as cultivars that, even in 1939, he faced up to the illogicality of the situation by listing "variants" where necessary, within the botanical varieties he felt compelled to use.

Each of the numbered trees was therefore a potential clone, so preparation for this present work has to a great extent been a matter of going through earlier lists, reappraising the horticultural value of each tree in the light of a further forty years of known behaviour, and deciding which of them justified its selection as a "New Cultivar".

SINCE 1939

Following a 1945 trip made by the author and his wife to survey and re-evaluate many of the hemlocks studied in 1938, an article entitled *A Swing around the Hemlock Circle* was published in the *American Nurseryman* (Swartley, 1946). This study also enabled a Supplement to the Thesis to be prepared in 1945 at the request of the Dutch nurseryman Peter den Ouden who was collecting information for a projected version, in English, of his book *Coniferen, Ephedra en Ginkgo* that had appeared in Dutch, in 1949.

This Supplement brought the Thesis of 1939 up to date. Several plants which had been passed by in 1938 because they were too immature to show individual characters now were considered worthy of naming. Examples are 'Wilton', 'Bristol', and 'Bradshaw'. Another, mentioned for the first time in the *American Nurseryman* article but overlooked in the Supplement was 'LaBar White-tip'. This is a beautiful fast-growing hemlock observed in the LaBar Nursery, Stroudsburg, Pennsylvania, and now widely planted.

The 1945 Supplement reflected the trend towards a better recognition of the essential difference between a minor botanical rank and a cultivar, although these latter were still listed as "varieties". Over the years that followed, additional selections were made by nurserymen and others, and the chosen names were published in one way or another. Most of these were picked up by the German writer Gerd Krüssmann in the 2nd edition of his work *Die Nadelgehölze* published in 1960, in which 23 garden forms are described.

The position finally was clarified when the 1961 edition of the *Cultivated Code* laid down that on or after 1st January 1959 a cultivar name must be a "fancy" one and that all cultivar names must be typographically distinguishable at a glance. The second edition of *Die Nadelgehölze* conformed to the new rule, as did also *Manual of Ornamental Conifers* by P. den Ouden and B. K. Boom which was published in 1965 and *Dwarf Conifers* by H. J. Welch which appeared the following year. Unfortunately P. den Ouden did not live to complete his book, but in preparing it for publication, Dr. Boom (with my consent) made full use of the Thesis of 1939 and the 1945 Supplement. Of seventy-two Canada hemlock cultivars listed and described by den Ouden and Boom, the descriptions of forty-seven were based on information supplied in this way by me.

Since neither paper had ever been published by Cornell University, the first "valid publication" of many of my names is the *Manual of Ornamental Conifers*. So if the authority for the name needs to be referred to, the correct form of citation (following the botanical pattern) is "Swartley ex den Ouden and Boom".

NOMENCLATURE TODAY

As recorded above, it had been the invariable custom to complete the name of each of the variant forms by adding one or more words (the ternary epithet) in Latin-form. In all or almost every case it is clear that—despite the often-inadequate description—the name arose from HORTICULTURAL SELECTION, RATHER THAN FROM BOTANICAL CLASSIFICATION. In present-day terms, the new plant was a cultivar. (This is often the case with horticultural writers even where the symbols "var." and "f." are introduced, since from the context it is clear that they are being used as abbreviations of the English words variety and form—or their equivalents in other languages. Thus Carrière lists his variants under the heading "*Variétés Horticoles*".)

As early as 1927, in an article in the *Journal of the Arnold Arboretum* **10** (1): 65, the late Professor Alfred Rehder had drawn attention to the clear distinction between botanical classification and horticultural selection. This he repeats in the Introduction to his *Bibliography of Cultivated Trees and Shrubs*, (1949) but it was not until 1953 that the separate needs of botanical and horticultural nomenclature were met by the publication of an *International Code for the Nomenclature of Cultivated Plants*. Consequently the practice of using Latin-form ternary names continued, together with the resulting confused thinking and some well-meaning attempts (now seen to have been an impossibility) to deal with what Rehder calls the "clonal garden forms" within the botanical system.

This then was the situation when the Thesis was in preparation: a botanical recognition and treatment of garden forms was still acceptable. It is now, of course, quite otherwise.

CLASSIFICATION: THE USE OF GROUPS

Apart from the limited attempts by Silva Tarouca and Beissner referred to already, no efforts had previously been made to produce a workable classification or grouping of the cultivars of *Tsuga*. In the 1939 Thesis, individual clones were classified according to foliage and habit. Twenty-two existing botanical varietal names were utilized. To accommodate the full range of variants in the material studied, five names were added. Three existing names were listed as "unclassifiable".

A type specimen was assigned to each accepted varietal epithet. If it was known to be a single clone, such as "var. *Hussii*," a plant of that name was treated as the "type specimen". If it embraced more than a single clone, now termed a clonotype, such as "var. *microphylla*," a typical clone was selected according to the best judgment of the author to serve as representative of the variety. In the Supplement of 1945, 116 variants were given major and 34 were given minor emphasis, all classified among twenty-four botanical varieties, as the term was at that time used.

An attempt has been made in the present book to establish a logical classification of hemlock cultivars, but the complexity of the material makes it difficult to develop an orderly and concise system, especially so in that the use of the infra-specific botanical categories for cultivars is no longer admissible. The study of hundreds of specimens fully demonstrates that variations of Canada hemlock are extremely diverse. One cultivar may combine two or three atypical characters: for example, long leaves, crowded leaves, or leaves not two-ranked (as in typical hemlock) are each sufficient to denote a distinct variation from the species. But all three of these characters are combined in certain individual clones. Furthermore, there are no separate and distinct limits on variability. Extreme complexity with complete gradation and intergradation is the rule in this genus. The lack of stability is another complicating characteristic of hemlock variations. This is especially true with plants that are slow-growing. Some cultivars that are rather dwarf and compact for many years will suddenly grow more rapidly. Examples are 'Doran', 'Cloud Prune', and 'Dwarf White-tip'.

So a great deal of cross-indexing has been necessary. Humphrey Welch's first book, *Dwarf Conifers* was a popular work, much of its value resting on the fact that he worked closely with Dr. Boom in its preparation. But in 1979 he followed it up with a definitive work *Manual of Dwarf Conifers*. For this book he had worked out a system of classification of cultivars based on the concept "Group" permitted by the *Cultivated Code*. Delays outside the present author's control in the publication of the present book have permitted an examination of Welch's system. I quote from the book:

> The *Cultivated Code* (Article 10, Note 2) states that the concept of a cultivar is essentially different from the concept of *varietas*. It could well have added the words 'and also of any other botanical category', since any such must always be a class of plants (the result of botanical classification) whereas a cultivar, whether it consists of a single clone or a thousand, is always a single entity seen through horticultural eyes (the result of horticultural selection). So although within a species the rank *forma* may as a general rule be regarded as the one most nearly approaching the concept of cultivar, it is still essentially different therefrom. The circumscription of a cultivar, however closely in any particular case at any particular time it may appear so to be, can therefore (because of this essential difference) never be the circumscription of a *forma* (nor of any botanical taxon) because, on the one hand, the latter concept must include individuals either existing or which may be found in the future which do not or will not fall within the circumscription of the cultivar, and because, on the other hand, a fresh selection giving rise to a new cultivar may at any time be made from within the *forma* (or as the case may be). It follows that any attempt to deal with the nomenclature of cultivars within the Botanical system is an impossibility. We cannot create a botanical combination out of a botanical non-entity. Such attempts may have succeeded in regimenting confusion: with the best of intentions they could never reduce it. As cultivars these plants began, cultivars they are now seen to be and (we must add) CULTIVARS THEY HAVE BEEN ALL THE TIME.
>
> The general trend amongst horticultural writers nowadays is to treat these names as cultivar names, but the pseudo-botanical combinations proposed by Professor Rehder (and others) are still widely regarded as valid, even if regrettable. But it now seems that any proposed botanical combination *based only on the description of a cultivar* does not satisfy the requirements of the *Botanical Code* regarding valid publication, the horticultural coiner of the name not having had the requisite formative intention (Welch, Taxon **27** (2,3): 187. 1978). It is therefore probable that this

pseudo-botanical status given to cultivars will eventually be pronounced illegitimate, so the plan followed in this book is to treat the original cultivar names as having had uninterrupted validity. Adoption of this plan has required a great deal of research to ascertain the intention of the writer who first used the name. But since, whatever their ultimate fate, these pseudo-botanical names are part of botanical history and will continue to be met with in reference books I have here included them in the synonymy of the names I treat as valid. Translation of these decisions into correct typography is simple. All I have to do with these pre-1st January 1959 Latin-form cultivar names is to see that my printer uses single quote marks and capital initials and all is in order.

Where a cultivar epithet so treated is derived from a personal or a place name such as 'Maxwellii' or 'Beuvronensis' no more be said, but if it consists of a descriptive adjective such as *pendula, fastigiata,* or *nana,* I have to consider, case by case, whether the description (although the name was given to a cultivar) also fits a CLASS of variants from the normal that is of sufficient botanical interest to justify recognition as a *forma.* But this is not, in any event, a matter that can be left to the printer. The cultivar named *Abies alba pendula* by Carrière did not (as it now seems) acquire botanical status or become a *forma* because someone once printed it *Abies alba* f. *pendula* Carr., and should I (or anyone) now feel it to be justified by the botanical circumstances in this or any other such case I would have to 'validly publish' the name as a new combination in compliance with the procedure laid down in the *Botanical Code,* supplying the so-far-missing botanical description, which in this case would have to include Carrière's plant along with all other variants—past, present or future—characterised by pendulous terminal growth. Were this ever to be done there would be a botanical taxon, *Abies alba* f. *pendula*—a class of plants—and falling within it three cultivars, Carrière's plant (this, the original clone, alone is entitled to bear the cultivar name 'Pendula'), another selection named 'Pendula Gracilis' by Sénéclauze and the new American clone 'Green Spiral'. Others might yet turn up!

But as a horticultural writer I have a much simpler course open to me.

In addition to the terms Clone and Cultivar already discussed, the *Cultivated Code* introduces another concept—the Group—a useful provision, since it introduces a valuable element of classification into this Code, in which groups are simply defined as 'Assemblies of similar cultivars'. Since the descriptive adjectives under discussion (*pendula, fastigiata, nana,* etc.,) often relate to groups of this kind, I can well afford to leave the unfortunate pseudo-botanical names to their fate and use these Latin epithets where I find it convenient as Group names under the *Cultivated Code.* Thus, '*Abies alba*—PENDULA GROUP' (containing the cvv. 'Pendula', 'Pendula Gracilis' and 'Green Spiral') is just as precise and useful (not to say, simpler) to horticultural readers than *Abies alba* f. *pendula* would be, so this simple solution of a very thorny problem is adopted here extensively.

The *Cultivated Code* does not require the use of any typography to distinguish a group name, merely suggesting that if used between the specific name and cultivar name it (the group name) be placed within brackets. This typography will doubtless serve very well for the complex cases it is provided for, but in the present and simpler case, I can as a grower and seller of conifers see practical objection to an arrangement that can give rise to an impression that two additional words have been introduced into the middle of the name. At best it would give my customers reasonable grounds for the complaint that we are 'Back again to the polynomials', at worst it would lead to unintelligible orders, extra work for nurserymen, disappointment for customers and frustration all round.

I therefore distinguish my group names throughout by the use of Capitals and Small Capitals, and where the group name is used in relation to the name of a particular cultivar I avoid using it between the specific name and the cultivar name.

This use of the Group category commended itself to the author, and since in the meantime he had been appointed Registrar by the Royal Horticultural Society acting as International Registration Authority for Ornamental Conifers it could be expected that his system would become general. So it was thought well to make use of the otherwise exasperating delay to re-cast the following section of this chapter: Mr. Welch having agreed to look it over before it went to press.

The system has two great advantages. Firstly, all the names appear alphabetically, making the finding of any particular name simple, and secondly, the complications arising from the somewhat arbitrary selection of the characters on which the various groups must be based, the numerous border-line cases and the logical claim of some cultivars to appear in more than one group can be overcome by the free use of cross-referencing.

Mr. Welch had consulted me with regard to the *Tsuga* cultivars and, indeed, drew most of his descriptions from the present book, which was even then in draft form. His latest book, therefore, agrees in the main with the present work, a few corrections that have proved necessary being noted in the descriptions later in this Chapter.

On his p. 375 Mr. Welch states "I have intentionally left proposals for any grouping of the cultivars to Dr. Swartley." This book is the grouping thus awaited. The group headings chosen again arrange themselves into groups. First come the foliage variations, next the colour forms and then the differences in habit. Lastly, selections of popular cultivars are listed according to their rate of growth—an attempt, to answer the unanswerable question "How big will it grow?"

Several of the variants in cultivation defy classification, so appear in no Group.

THE CULTIVAR GROUPS

The first eight groups are based upon foliage deviations from the normal.

LITTLE-LEAF GROUP

Tsuga canadensis f. **microphylla** (Lindl.) Beissn. Syst. Eintheil. Conif. 40. 1887; (*Abies canadensis microphylla* Lindley, Gard. Chron. **20:** 460. 1864) *Plants differing from the normal by having leaves not more than 8 mm long.*

Plants distinguished by having exceptionally short leaves turn up occasionally in the seed-beds or are found in the wild and at least one is recorded as having arisen as a bud-mutation. Some of these have leaves no longer than 6 mm. They are recognised botanically as f. *microphylla*. This word would itself have served well as a Group name, but its English translation may be more readily understood by users of this book. In such circumstances the use of the basionym as a cultivar name—'Microphylla'—should be restricted to the clone described by Lindley, should this ever be identified (which after so long passage of time is unlikely). All other clones selected for cultivation should be registered under new clonal names.

THE SAME LIMITATION ON THE USE AS A CULTIVAR NAME OF A BASIONYM GIVEN VALID BOTANICAL RECOGNITION SHOULD BE ACCEPTED IN EVERY CASE.

'Beehive'	'Microphylla'
'Brevifolia'	'Milfordensis'
'Bristol'	'Parvifolia'
'Jenkinsii'	'Salicifolia'
'Kingston Hollow'	'Slenderella'

LARGE-LEAF GROUP

Tsuga canadensis f. **macrophylla** (Beissn.) Swartley, Grad. Nov.; (*Tsuga mertensiana macro-phylla* Beissn. Handb. Nadel. 404. 1891). *Plants differing from the normal by having leaves exceptionally large, up to 18 mm in length.*

These are hemlocks that have exceptionally large leaves, the longest measuring 18-21 mm as compared with 14 (to occasionally 16) mm for typical hemlock. Most LARGE-LEAF hemlocks are of slow growth, dense and compact, with closely set leaves and branch-structure heavier than normal.

'Atrovirens'
'Broughton'
'Curtis Ideal' See CONICAL GROUP
'Dover'

'Hicks'
'Jeddeloh' See SPREADING GROUP
'Macrophylla' See Note under
   'Little-Leaf Group
'Towson'

SPARSE-LEAF GROUP

Tsuga canadensis f. **sparsifolia** Beissn. Handbuch, Nadel. 402. 1891. *Plants differing from the normal by bearing leaves distant along the shoot and by an irregular arrangement of the leaves on the shoot.*

Although not an uncommon deviation, this is not a good characteristic upon which to base a group, since it tends to disappear as the plant ages. The plants in this group present a sparse appearance, usually from two different aberrations. There are fewer needles per centimeter of branchlet than on typical hemlock, and also their arrangement is quite irregular. In the Hemlock Thesis the author recorded several plants that at the time appeared to belong to this group. This included plants from Wildacre Nurseries, Lenoir, North Carolina; W. I. Valentine and Sons, Cosby, Tennessee; Blue Ridge Nurseries, Harrisburg, Pennsylvania; Curtis Nurseries, Callicoon, New York, and the F. and F. Nurseries, Springfield, New Jersey. Some of these selections have subsequently disappeared from cultivation. Others have outgrown the sparse-leaf character, which may be one feature of juvenility.

Two cultivars, 'Valentine's Yewlike' and 'Jennings Yewlike' recorded in this group in 1939, have been reclassified and put into the YEW-LIKE GROUP.

'Jenning's Yewlike' See YEW-LIKE GROUP
'Sparsifolia' See Note under LITTLE-LEAF GROUP
'Valentine's Yew-like' See YEW-LIKE GROUP

DENSE-LEAF GROUP

Tsuga canadensis f. **densifolia** (Jenkins) Swartley. Grad. Nov.; (*T. canadensis* var. *densifolia* Jenkins, Heml. Arb. Bull. 6. 1934). *Plants differing from the normal by leaves noticeably more densely set.*

Dense-Leaf variations of Canada hemlock are quite common. They occur frequently both in the wild and in nursery rows. The group is more or less a catch-all, and the inclusion of certain cultivars and the exclusion of others is inevitably arbitrary. These trees are mostly of slower growth than normal hemlock and lacking in grace, with supraplanate leaves that are conspicuously crowded (averaging at least 20 per cm. of branchlet) but normal in length.

'Abbott's Dwarf'   'Muttontown'
'Bacon Cristate'   'Redding'
'Ehrle Largeleaf' See LARGE-LEAF GROUP   'Rockland'
'Everitt Golden' See GOLDEN GROUP   'Von Helm'
'Great Lakes'   'Waverly'
'LaLar Gem'   'Wheelerville' See
'Lewis'   CONICAL GROUP

WIDE-LEAF GROUP

Tsuga canadensis f. **latifolia** (Sén.) Swartley. Grad. Nov.; (*Tsuga canadensis latifolia* Sén., Conif. 19. 1868). *Plants differing from the normal by bearing leaves noticeably wider than normal, relative to their length.*

A group of trees of less graceful habit than normal hemlock with shorter and very much broader, glossy leaves (the length averaging not more than five times the width). The plants can be mistaken for *T. sieboldii.*

During studies for the 1939 Thesis, a number of plants of this type were examined and recorded. Because the epithet *latifolia* is accepted for a minor botanical classification, and since the plant or plants described by Sénéclauze in 1868 cannot now be identified, it cannot now be used as a cultivar name, so new names acceptable under the Code have had to be found.

'Latifolia' See Note under LITTLE-LEAF GROUP
'Dawsoniana'
'Heli'

CINNAMON-TIP GROUP

This is a gardeners' classification, not recognised botanically. These plants are generally broad, globose dwarfs with the ends of the twigs noticeably swollen and thickly covered with brown tomentum. A branch spray consists of seven or eight branchlets resembling a fan. The needles are usually narrower than typical hemlock, sometimes spoon-shaped at the apex. Variants of this group are certainly regarded as oddities, although several cultivars are not unattractive. It is a rather common variation.

'Andrews'
'Bergman's Gem
'Cinnamomea'
'Cushion'
'Drake'
'Palomino'
'Rugg's Washington Dwarf'
'Stewart's Gem'
'Van Dyne'

TWIGGY GROUP

This is another gardeners' classification that includes slow-growing shrubs or small trees with very crowded twigs and radially disposed, noticeably short and crowded leaves. One characteristic which is common to the different cultivars of this group and contributes to the crowding of the twigs, is the closeness of the buds. On some specimens as many as a dozen buds are clustered in an area about 3 mm in diameter. Plants of 'Hussii' have been successfully used in bonsai training.

This group raises problems of identification, since the various clones are very similar in appearance. The distinction is usually more in habit than in leaf structure, so it is often impossible to identify a clone from a specimen of foliage alone, without a knowledge of the plant from which it was taken.

Considerable confusion has been produced by the late Henry Hohman having widely distributed a clone (here re-named in his memory) in this group under the incorrect name 'Gracilis'.

'Abbott's Pygmy'	'Horsford'
'Armistice' See CONICAL GROUP	'Hussii'
'Coffin'	'Jervis'
'Connecticut Turnpike'	'Lewis'
'Gracilis' SENSU Hohman. See 'Henry Hohman'	'Little Joe'
'Greenwood Lake'	'Minuta'
'Hancock'	'Rhapsody'
'Henry Hohman'	

YEW-LIKE GROUP

A minor deviation from the normal that is difficult to define and unstable and which certainly does not merit botanical recognition. The plants in this group are all slow-growing and with foliage somewhat yew-like in appearance; leaves crowded at the ends of each year's growth, very irregular in arrangement and varying considerably in length. The Yew-like character is, however, due mainly to the general impression given by the plant as a whole rather than to the detail of the foliage, and along with the tendency of all variations noticeable in young plants it is apt to lessen and even disappear as the tree ages.

'Jenning's Yew-like'
'Taxifolia'
'Valentine Yew-like'

The two groups that follow are colour forms:

WHITE-TIP GROUP

Tsuga canadensis f. argentea Nelson, Pinaceae. 32. 1866. (*Abies canadensis* f. *alba-spica* Barron ex Gordon, Pinetum ed. 2, 421. 1875; *Tsuga canadensis* f. *albo-spica* Beissn. Eintheil. Conif. Benennung. 40. 1887. (Name only); *Tsuga canadensis* fol. *argent. varieg.* Hort. ex Beissn., Handbuch. Nadel. 403. 1891). *Plants having the tips of the current year's growth white or nearly so in mid-summer.*

The earliest recorded botanical epithet is *argentea,* but 'Albo-spica' is often loosely and inaccurately used for any clone with this character. Although pseudo-botanical names (as above, with several alternative spellings to be found) have been given to plants with more or less white young growth, in each case it is clear that what is being described is a cultivar and always the descriptions are insufficient to identify the clones in cultivation today, with the exception of 'Albo-spica' which may reasonably be given the benefit of the doubt because of its recorded European origin.

Most clones in this group have a common characteristic, which is that the whitish color appears with the second or summer growth on the tips of the twigs and the short lateral shoots. 'Frosty', which has a general overall whiteness in shade and a dirty color in sun, 'Watnong Star' and 'Betty Rose' are exceptions to the rule in that the white coloration appears in Spring. The color in 'Dwarf White-tip' and 'Imperial' tends to fade by autumn, but in 'Snowflake' and 'Gentsch White' it holds into the winter. The intensity and duration of color varies, depending not only on the cultivar but also on the plant's environment, including exposure to sunlight and nutrition. Regular heavy pruning accentuates the effect.

Many plantsmen and gardeners dislike plants with white or yellow leaves because they appear to be sick. This is mere prejudice, for there is usually ample chlorophyll in the green parts of the leaf to carry on normal photosynthesis, and a well-grown specimen can make a quite spectacular garden plant.

'Albospica'	'Gentsch Variegated' See 'Gentsch White'
'Argentea'	'Imperial'
'Bergman's Frosty' See 'Frosty'	'LaBar White-tip'
'Betty Rose'	'Raraflora Snowflake' See 'Snowflake'
'Creamey'	'Silver Tip'
'Dwarf White-tip'	'Snowflake'
'Far Country' (J.210)	'Summer Snow'
'Frosty'	'Watnong Star'
'Gentsch White'	

GOLDEN GROUP

Tsuga canadensis f. **aurea** (Nelson) Jäger u Beissn. Ziergehölze. 495. 1884. (*Abies canadensis* var. *aurea* Nelson, Pinaceae. 32. 1866.) *Plants distinguished by foliage that is yellow (or definitely yellowish) during at least part of the year.*

The oldest name in this group is 'Aurea'. According to Beissner (1891) var. *aurea* is "slow growing with gold-yellow branch tips, really decorative," but Webster (1896) stated that var. *aurea* is "no great acquisition, the coloring being both irregular and inconsistent." Many different clones have been called var. *aurea* down through the years so the name should not now be used in any clonal sense. Most of them are really a transient yellow-green and the author is inclined to agree with Webster that for most of the year the coloration is a sickly green. A "golden" hemlock from Brimfield Gardens Nursery, Wethersfield, Connecticut, called 'Brimfield Golden', is scarcely worth planting. There are however, Dense-Leaf variants, all rather similar, that display a more definitely golden coloration, especially during certain parts of the year.

'Aurea'
'Brimfield Golden' See note above.
'Everitt Golden'
'Silvery Gold'

The following five groups are based on habit of growth.

GLOBOSE GROUP

Tsuga canadensis f. **globosa** Beissn. *Handbuch Coniferen-Benennung 65. 1887. Plants characteristically forming symmetrical, rounded bushes or small trees, usually with no leader but, instead, with multiple stems of approximately equal length; having a width more or less equal to the height.*

Although the main characteristic of this group is clear, the toleration allowable must be arbitrarily settled. In the Hemlock Thesis, this limit was taken as a disparity between the height and width of the plant of not more than 25% either way. There are, of course, the inevitable border-line cases, and cultivars listed under foliage groups may also be globose in shape. This is true of most cultivars in the CINNAMOMEA GROUP.

A difficulty experienced in this group is the tendency of all habit variations to decline or even disappear with age. Some plants may remain globose for twenty-five years, then suddenly shoot upward. Some cultivars, noticeably 'Geneva' are liable to lose character if grafted, due presumably to the influence of the understock.

'Andrews'	'Laurie'
'Bagatelle'	'Mansfield'
'Beehive'	'Matthews'
'Bristol'	'Rock Creek'
'Far Country' (J.210)	'Unique'
'Geneva'	'Warner's Globe'
'Innesfree'	'Wilton Globe'

CONICAL GROUP

Tsuga canadensis f. **pyramidalis** Jenkins, Hemlock Arboretum Bulletin 6. 1934. *Plants that develop a noticeably tapered outline, intermediate between the round-topped tree characteristic of the species and a fastigiate habit of growth.*

Variations in habit are difficult to classify because of changes that occur as a tree matures. It is a classification of extremely doubtful botanical value. There is no guarantee that a twenty-five-year-old plant will be a larger replica of a ten-year-old plant. There is no cultivar that can be called typical of this group but most are good compact and slow-growing forms.

'Armistice'	'Guldeman's Dwarf'
'Baldwin Dwarf Pyramid'	'Harmon'
'Boulevard'	'Hiti'
'Bradshaw'	'Hornbeck'
'Brandley'	'Jacqueline Veckade'
'Curtis Ideal'	'LaBar Gem'
'Doc's Choice'	'Meyers'
'Doran'	'Moll'
'Essex'	'Popeleski'
'Fremdii'	'Pumila'
'Geneva'	'Rockport'

FASTIGIATE GROUP

Tsuga canadensis f. **fastigiata** Beissn. Handbuch Nadel. 402. 1891; *T. c. f. columnaris* (Bolle) Beissn. loc. cit. *Plants developing a narrow, upright habit, having a height more than twice the breadth.*

This group is presented without apology. It is more of a gardeners' classification than a botanical one, and as already indicated, the relative proportions and habit of any hemlock cultivar can change with maturity of an individual plant or be affected by the cultural treatment it receives, especially as a young plant. So more study and experience is needed before we can accurately predict the growth habits of some of these plants. Two clones were listed and described in the same year, namely 'Fastigiata' and 'Columnaris'.

'Columnaris'	'Moon's Columnar'
'Fastigiata'	'Schramm'
'Kingsville'	'Vermuelen's Pyramid'

SPREADING GROUP

Tsuga canadensis f. **prostrata** *Plants without any central stem and with a horizontal or wide-spreading branch system, resulting in a flat-topped plant much wider than high.*

This again is a somewhat arbitrarily chosen characteristic. Over the years numerous spreading clones have been selected and named. Some of them are strictly prostrate, but the border-line between this and the WEEPING GROUP is almost impossible to determine, since the tendency of the terminal growths to hang down that is a characteristic of the species (and one to which it owes so much of its beauty) is often more or less a noticeable feature also of these spreading forms. Some are very decorative. Despite all this, many eminent plantsmen are still unaware of the existence of such plants, forms that should have great potential value, especially in the home landscape.
The use of the name f. *prostrata* by Bean (Trees and Shrubs ed. 1, **3**: 482, 1933) was a case of misidentification. (See p. 66.)

'Barrie Bergman'	'Jan Verkade'
'Beaujean'	'Jeddeloh'
'Bennett'	'John Swartley'
'Bonnie Bergman'	'Kathryn Verkade'
'Cloudprune'	'Minima'
'Cole's Prostrate'	'Narragansett'
'Corbit'	'Nana'
'Fantana'	'Outpost'
'Gracilis' (NON Hohman)	'Saratoga Broom' See 'Beaujean'
'Greenspray'	'Sherwood Compact'
'Hahn'	

PENDULA GROUP

Tsuga canadensis f. **pendula** Jäger u Beissn, Ziergeholze ed. 2, 445. 1884. *Plants of more or less normal habit with ascending main branches or with a defined trunk, but distinguished by having all their terminal growths markedly pendulous.*

Although all early references to the "Sargent" hemlock refer to it as a cultivar, the epithet *pendula* was also used uncritically and interchangeably by American writers. A pendulous form of Canada hemlock was known also in Europe, and at an early date this was given botanical recognition by the German writers Jäger u Beissner. Bean (*Trees and Shrubs Hardy in the British Isles,* Vol. II, 606. 1914) was the first writer to record the difference in character of the two forms. Of his 'var. *pendula*' he wrote: "A very attractive shrub or small tree forming a hemi-spherical mass of pendulous branches, completely hiding the interior" whilst 'var. *Sargentii*' he describes as "Another pendulous form of more compact shape than the preceding. Found about 1870 in the Fishkill Mountains, New York".

The reference by Michaux in 1803 (See p. 63) to the propensity of the hemlock spruce to produce variants of this nature is clear evidence that *Tsuga canadensis* f. *pendula* must include plants not originating with the Sargent findings.

Faced with this distinction, it is best to follow Welch (*Manual of Dwarf Conifers,* 1979) in accepting Beissner's botanical classification, adopt his epithet as our Group name and restrict the use of 'Pendula' as a cultivar name to the European form. There are now several selections in cultivation in the PENDULA GROUP.

'Abbott Weeping'	'Green Cascade'
'Ashfield Weeper'	'Horton'
'Brookline'	'Hunnewell'
'Callicoon'	'Kelsey's Weeping'
'Cappy's Choice'	'Lustgarten Creeping'
'Cesarini's Broom' (Provisional)	'Outpost'
'Curtis Weeping' (Provisional)	'Pendula'
'David Verkade'	'Sargentii'
'Detmer's Weeper'	'Starker'
'Elm City'	'Valentine'
'Gable Weeping'	'Wodenethe'

GROWTH POTENTIAL

Finally, we come to a gardeners' classification: a listing of the garden cultivars according to their vigour. It is perhaps the most difficult task of all—an attempt to answer the invariable question "How big will it grow?" There can never be an answer to this apparently simple question that is not linked with time. An assessment of a plant's capacity for growth that will be spot-on at ten years would be widely out fifty years later. The following listings must therefore be used with caution. The listings assume normal climatic and environmental conditions and are estimates of probable size after 10-15 years growth, and of course these listings by size cut across the Groups.

COMPACT FORMS

There are variants noticeably more compact than typical Canadian hemlock and often little different therefrom in foliage character or habit. They are often attractive garden plants for a number of years, but they all are potentially small to medium-size trees.

'Albo-spica'	'Brandley'	'Dawsoniana'
'Angustifolia'	'Bristol'	'Doc's Choice'
'Aurea'	'Brookline'	'Fremdii'
'Bagatelle'	'Broughton'	'Gable Weeping'
'Boulevard'	'Coplen'	

'Geneva'   'Microphylla'
'Great Lakes'   'Moll'
'Harmon'   'Nana'
'Jenkinsii'   'Pendula'
'Hiti'   'Starker'
'Kelsey's Weeping'   'Stockman's Compact'
'LaBar Gem'   'Stranger'
'LaBar White-Tip'   'Taxifolia'
'Laurie'   'Wilton'
'Matthews'   'Wodenethe'
'Meyers'   'Youngcone'

DWARF FORMS

These are much slower-growing plants, not likely to reach eye-level or become more than of shrub-size for many years. It is therefore a very useful range of garden plants, especially where space is at a premium.

'Abbott's Dwarf'   'Dwarf White-Tip'   'Jeddeloh'
'Armistice'   'Elm City'   'Jervis'
'Bacon Cristate'   'Everitt Golden'   'Kathryn Verkade'
'Beaujean'   'Fantana'   'Kingston Hollow'
'Beehive'   'Far Country' (J.210)   'Minima'
'Bennett'   'Gable Weeping'   'Nearing'
'Bonnie Bergman'   'Gentsch White'   'Palomino'
'Callicoon'   'Gracilis'(NON Hohman)   'Starker'
'Cinnamomea (the whole Group)   'Green Cascade'   'Stockman's Dwarf'
'Cloudprune'   'Greenspray'   'Valentine'
'Corbit'   'Greenwood Lake'   'Verkade Recurved'
'Creamey'   'Hahn'   'Von Helms'
'Curly'   'Harmon'   'Warner's Globe'
'Curtis Ideal'   'Heli'   'Watnong Star'
'Curtis Spreader'   'Henry Hohman'
'Doran'

EXTREME DWARF FORMS

There are diminutive plants suitable for specialized collections, rock gardens, and bonsai purposes. Canada hemlock has produced a surprisingly large number of these diminutive plants growing only 1 inch or less per year.

'Abbott's Pygmy'   'Helene Bergman'
'Andrews'   'Hornbeck'
'Bacon Cristate'   'Horsford Compact'
'Barrie Bergman'   'Hussii'
'Bergman's Gem'   'Minuta'
'Betty Rose'   'Pygmaea'
'Coffin'   'Rhapsody'
'Cole's Prostrate'   'Rugg's Washington Dwarf'
'Essex'   'Unique'
'Hancock'   'Verkade Petite'

DESCRIPTIONS OF THE CULTIVARS OF TSUGA CANADENSIS

The remainder of this Chapter consists of an alphabetical listing of all the names I have found, whether past or present (stopping short of an exhaustive search) in living collections and nursery catalogues and libraries in the United States and Great Britain. There is no doubt that this list is already far too long. Many of the cultivars are insufficiently distinctive—to the point that some cannot be told apart even by the experts or by those whose livelihood depends on their propagation.

T. canadensis is very slow-growing in its early years and young plants often belie their capacity for stronger growth in maturity and have an attractiveness (as do other young things such as puppies and kittens) which they eventually lose. It is well recognised in many genera that plants will wholly or partly grow out of any peculiarities they showed when young, and in result the mother plants of a number of the older cultivars now show little or no difference from the typical Canadian hemlock. Propagations may revive the garden attractiveness of their mother for a while, until they too lose this early promise.

This raises the probability that many of the more recent selections will in time go the same way. As with many other conifer cultivars that are regarded as dwarf (when in fact they are merely slow-growing), those forms only justify recognition because of their garden value for a limited number of years. In other words, they must be thought of as having a limited useful life and be planted accordingly, rather than being permanently set out for the benefit of posterity.

It follows therefore that restraint, much more than has been the case in the past, should be exercised in the selection and "naming" of new cultivars. This process is governed by the *International Code of Nomenclature for Cultivated Plants* (usually referred to as the *Cultivated Code* or ICNCP).

This *Code* governs how you shall do it but not what you shall do! Provided you follow the rules you can "validly publish" new cultivar names galore—lengthening peoples' lists but adding no beauty to their gardens. The proliferation of cultivar (clonal) names for Canadian hemlocks is a particularly bad example of this. (In general, in the case of long-lived trees such as conifers seldom if ever propagated from seed, a cultivar will consist of a single clone. So, but only as a special case, these terms are often interchangeable in this particular context.)

The *Code* applies to all groups of garden plants, but it provides also in appropriate cases for the establishment of *International Registration Authorities* to exercise a more detailed control of a particular group. For some years, until about 1977, the Arnold Arboretum acted as *National Registration Authority*, but the *International Registration Authority for Ornamental Conifers* is now the Royal Horticultural Society, London S.W.1.

The existence of an International Registration Authority for a particular group of plants indicates that a need is felt in that group for more detailed guidance than is afforded by the *Cultivar Code*. For instance, in the case of conifers it is understood that the Registration Authority has laid down that in no circumstances shall a cultivar name be re-used even where the original clone is lost to cultivation, and also that no name shall be repeated within a genus. (Thus if there was a *Tsuga canadensis* 'Polly', 'Polly' would not be an acceptable name in, say, *Tsuga mertensiana*). The Authority has also laid down guide-lines for choosing cultivar names that are intended to avoid the risk of confusion in future; hence the value of a preliminary approach to the Registrar by anyone wishing to register a new cultivar name.

This might be a good point at which to refer to certain types of name that are of undesirable form and likely to cause confusion. Such should, of course, be avoided when coining names for new introductions.

A cultivar name is an arbitrary device for identifying a particular cultivar. It must consist of a fixed arrangement of symbols (numerals or the letters of some recognised alphabet) which can be recorded in writing (as one, two or at a pinch three words in some living language) and be capable of being easily pronounced.* This makes the number of possibilities almost inexhaustible but clearly other considerations must come into view. Any name liable to cause confusion is ruled out under the Code; one that transgresses the laws of common-sense and good taste must also be rejected. Any thing that is intended for use in everyday life should be pleasant and easy to use, without strain on the memory, and this applies to cultivar names. Even so, the range of suitable names is enormous.

No objection normally arises from the very frequent practice of commemorating the place of origin or the honouring of a person or a person's memory, but it is not the function of a name to form part of even a rudimentary system of classification, and any such attempt is therefore undesirable. A number of such names (e.g. 'Will's Zwerg', 'Gable Weeper', 'Bloomer's Dark Globe') are well established in use so must be retained, but any increase in this type of name will inevitably result in confusion. If five nurserymen, Adams, Brookes, Charles, Dean and Edwardes each have a 'Little-leaf', a 'Large-leaf', a 'Sparse-leaf', a 'Spreader', a 'Weeper' and a 'Twiggy', there will thirty different cultivars. A customer carelessly asking for a plant of 'Adams' may get an 'Adam's Little-leaf', an 'Adam's Large-leaf', and 'Adam's Sparse-leaf', and 'Adam's Spreader', an 'Adam's Weeper' or an 'Adam's Twiggy', and an order for 'one of the weeping ones I saw on the nursery' may be filled by supplying any one of five different cultivars.

Another source of almost certain trouble is to coin a new name consisting of an old one, plus an extra word. (If there is already a cultivar named 'Boston' do not introduce 'Boston Gem' or 'City of Boston.')

Names that have never been "validly published" according to the rules in the *Cultivated Code* have no status, however widely they have been used in the trade. Where I have been able to secure some agreement to changing an undesirable name, or where a more acceptable name suggests itself I have used it. In all other cases, as also in the case of names undocumented or insufficiently known, I have listed the name as 'Provisional'. This gives the name (for what value it has) its alphabetical place, but does not constitute valid "publication". It leaves it open for a more acceptable name to be chosen and registered in the future.

USE OF TYPOGRAPHY

In the following lists accepted names are introduced by the use of bold type, - - **'Andrews'** (The hyphens stand for the words *Tsuga canadensis,* or as the case may be.) They are in the following categories.

Registered name. A name registered at Arnold Arboretum before 1977 or by the International Registration Authority since that date.

New cultivar. A name being (so far as is known) validly published here for the first time.

New Name. A name being validly published in this book to replace an illegitimate or unacceptable name (e.g. 'Henry Hohman'.)

Unaccepted names are introduced in normal type.

Synonyms and cross references.

Provisional names. This category, already mentioned, is one permitted by Article 34 of the *Botanical Code.* Under that code, (and by analogy when used in a horticultural context) such names are not validly published—yet they provide a convenient way of identifying a clone meanwhile for purposes of argument or enquiry, pending a

---

*The actual wording of the Code is that a cultivar name nowadays must be a fancy name, that is, not a botanical name in Latin form. Since the objective is to make the cultivar part of the name to stand out at a glance from the botanical part it is wisest to implement this rather curious wording by avoiding altogether any word that has a look of Latin about it.

decision later on as to its claim for recognition as a new cultivar when full information becomes available. It is a useful temporary resting-place for all the undocumented, uncertain and little-known names that there will always be, but all *Provisional names* should eventually be cleared up, one by one, and the finding of more acceptable names or the coming to light of full information should justify up-grading of many of the *Provisional names* in this edition to *Accepted names* in a later one.

After each accepted name will be found whatever information is reliably available. The amount varies considerably from case to case. The target has been to record the origin of the cultivar and a reference to the first use of the name (this is usually also its "valid publication"), together with any other bibliographical citations considered useful in a particular case; an appraisal of the garden-value of the plant and a non-technical description to help its identification and separation from similar clones. In this matter, the liberal use that has been made of photographs will be of help.

In many cases, the available information falls well short of the target. Correspondence (which, on account of the author's present state of health should be addressed to him "c/o The Publisher") is invited from anyone able to confirm, amplify or correct the information recorded for any cultivar or to draw attention to newly introduced cultivars.

Tsuga canadensis 'Abbott Weeping' Krüssman, Nadel. 341. 1972; NON Welch, Dwarf Conif. 320. 1966. PENDULA GROUP. This was found in Smith's pasture near the road not far from Athens, Vermont. In 1938 it was about 16 feet × 16 feet, a small bush tree with two trunks and very irregularly pendulous branches. In 1945 the author stated that this cultivar was intermediate between 'Sargentii' and 'Outpost'. Stout (New York Bot. Gar. Journ. **40:** 153. 1939) quoted a letter from Frank Abbott to the effect that this variation is more like 'Kelsey's Weeping' than any of the Sargent hemlocks. The use of this name by Welch (Dwarf conif. 320. 1966) was clearly a mistaken reference to 'Abbott's Pygmy'.

T. canadensis 'Abbott Weeping'. Near Athens, Vermont, in 1938. Discovered by Frank L. Abbott.

- - **'Abbott's Cinnamomea'** Hort. See 'Cinnamomea'.
- - **'Abbott's Dwarf'** Hillier, Dwarf Conif. 65. 1964; NON Welch, Dwarf Conif. (pro. syn.) 320. 1966. DENSE-LEAF GROUP. As seen at Kingsville Nurseries, Kingsville, Maryland, this was a slow-growing, conical bush 75 cm × 50 cm, densely furnished right through the plant with tightly packed, dark green foliage and long (up to 15 mm) and wide leaves except just at the tip of the shoot and along its upper side, where short (to 5 mm) leaves lie forwards along the shoot at 10° and twisted so as to expose the white stomata lines, giving an interesting two-tone effect to the plant.

  A plant in the U.S. National Arboretum, Washington, D.C. (Accession No. 20646), now 30 years old, is a squat, irregular bush 5 feet high × 6 feet wide with strongly ascending branches having their tips slightly pendulous. The leaf sprays are flattened, needles greyish-green. Another tree in the same collection (Accession No. 16131) acquired from Henry Hohman at Kingsville Nursery in 1960 is similar, but 9 feet high × 14 feet across. In this older tree the lower branches are horizontal, and the upper branches still more steeply ascending, but still arching over at their tips.
- - 'Abbott's Fountain' Provisional name. Listed by Hohman, Kingsville Nurs. Cat. c. 1970, and in some present-day collections. It forms a large, spreading, open-centered bush with normal leaves. A better name should be registered.
- - **'Abbott's Pygmy'** Jenkins, Heml. Arb. Bull. 62: 1948; 'Pygmaea' Spingarn, Amer. Rock Gar. Soc. Bull. **27**: 89. 1969, Welch, Dwarf Conif. 321. 1966. TWIGGY GROUP. This is probably one of the smallest hemlocks, forming an irregular globe with leaves shorter and annual growth less than 'Minuta.' The leaves are small, seldom over 5 mm by 1 mm, elliptic, pointed, very dense and irregular, and mid to light green. The shoots are light brown. The buds are relatively large and consequently conspicuous. It was collected by Frank L. Abott in May 1933, north of Richmond, Chittenden County, Vermont, on the west side of the Winooski River, on the same ridge as 'Minuta'.
- - 'Albo-picta' Hort. See 'Albo-spica'.
- - **'Albo-spica'** Jäger u Beissn. Ziergeh. ed. 2, 445. 1884. (spelling *alba-spica*); *albo-spicata* Beissn. Mitt. d.d.d. Ges.**15**: 151. 1906. WHITE-TIP GROUP. It was described as a compact form with whitish branch tips which are most intense in midsummer. Since this plant at Far Country Hemlock Arboretum had been received from Hillier & Son, Nurserymen at Winchester, England, it seems reasonable to identify it as the Jäger u. Beissner plant. The branchlets are more crowded than normal hemlock, which otherwise (save for the colour) they closely resemble. The spelling 'Albo-spica' is now general, but the name should not be applied to other clones such as 'LaBar White-Tip'.
- - 'Albo-spicata Hort. See 'Albo-spica'.
- - 'Amawalk' Provisional name for a clone probably no longer in circulation.
- - 'Albo-spicata Imperial' (Raraflora Nurs. Cat. c 1970) See 'Imperial White-tip.'
- - **'Andrews'** New Name. Jenkins, Heml. Arb. Bull. 15. 1936. CINNAMON-TIP GROUP. Jenkins records the receipt of a small plant from Howard S. Andrews, a nurseryman at Seattle, Washington under the name var. *minuta-nana*. Although this is a variant that is extremely slow-growing it has no connection with 'Minuta' and the clear intention of Recommendation 12A of ICNCP justifies rejection of a name that "could cause confusion". This cultivar has been observed in the collection of Joel Spingarn, Baldwin, Long Island, New York.
- - **'Angustifolia'** Swartley ex den Ouden and Boom, Man. Conif. 450. 1965. A loosely compact plant, moderate in growth, with tufted branchlets; lateral branchlets tend to be as long as the terminal; twigs heavier than typical hemlock with extremely narrow leaves obtuse to acute at the apex, sparsely toothed. The mother plant was presented to Far Country in 1942 by George L. Ehrle, Clifton, New Jersey. At that time was 6 feet high and a little more in width. In 1970 it was 16 feet high about 20 feet wide. Although the Swartley Thesis of 1939 (not having been published) does not legitimise the name, the reference (Hemlock Arboretum Bulletin 39) in 1942 to "var. *augustifolia*" can be accepted as valid publication, despite the obvious mistake in the spelling.

T. canadensis 'Abbott's Pygmy'. Kingsville Nursery. Photograph taken in 1970.

T. canadensis 'Angustifolia'. U.S. National Arboretum, Washington, D.C. Accession No. NA 21183.

T. canadensis 'Armistice'. Type 1, conical. U.S. National Arboretum, Washington, D.C. Photograph taken in 1970.

T. canadensis 'Armistice'. Type 2, flat-topped form. U.S. National Arboretum, Washington, D.C. Accession No. 20834.

Foliage detail of 'Armistice'.

- - f. **argentea** See p. 73.
- - **'Argentea'** Nelson, Pinaceae 32. 1866. (*Abies canadensis argentea*) "The silver variegated" form. This clone cannot now be identified, so the name should not be used.
- - **'Argenteovariegata'** Hort. ex Beissn. Handbuch Nadel. ed 2. 90. 1909. See 'Variegata'
- - **'Armistice'** Warner ex Spingarn, Amer. Hort. Mag. **44**(2): 99. 1965; Amer. Rock Gar. Soc. Bull. **27**(3): 86. 1969. (In both cases with a dubious description.) CONICAL GROUP. This was a selection made by the late Ralph M. Warner, Milford, Connecticut. It normally makes a rather open bush with growth in rather strong, noticeably flat sprays and branchlets very close together with foliage very dense. The leaves are oval, broad and very densely set, forming a spray that is almost solid leaf. Leaves are shiny dark green with two narrow bands of stomata on the underside.

  A plant in the Gotelli collection at U.S. National Arboretum, Washington D.C. (Accession No. 20836) is on record as having been purchased by Mr. Gotelli from Warner in 1957. It is now an upright plant of conical outline 6½ feet high by 5 feet wide, with horizontal branches with upturned tips and short, stiff branchlets. The needles are dark green. The author observed several other plants in the same collection. Most were pyramidal, very compact; one in particular was flat-topped and irregular. About a dozen were seen in the nursery of Rudolph Kluis, varying considerably in habit.

  There is some mystery here, since plants under this name show considerable variation in habit. The explanation might be that Ralph Warner sent out two clones which in small sizes in his nursery were indistinguishable, or it might be that the mutation that gave rise to the cultivar was not quite stable and that, in result, the selection of propagating material influences the resulting plant. This phenomenon is not unknown, *Picea abies* 'Maxwellii' being a well-known example.

  I suggest, provisionally, under 'Muttontown' a way of dealing with this difficulty.
- - **'Ashfield Weeper'** New Cultivar. Listed by Watnong Nursery. 6. 1967. (Name only.) PENDULA GROUP. This is an irregular weeper that starts out like 'Kelsey's Weeping' but is faster growing and less irregular when it gets older. The original plant was discovered along the roadside in Ashfield, Massachusetts, in 1964 by Ormand Hamilton of Worcester, Massachusetts. He sent propagating material to the U.S. National Arboretum and they sent a rooted cutting to the Arnold Arboretum in 1967. A plant, Accession No. 377-64, is growing there.
- - **'Atrovirens'** Beissn. Mitt. d.d.d. Ges. **6**: 89. 1897; *Abies canadensis atrovirens* Parsons, Nurs. Cat. 1890. LARGE-LEAF GROUP. Beissner described this as a medium-compact form with conspicuously dark-green leaves. The description in a catalogue issued in 1890 by the one-time famous nursery of S. B. Parsons & Sons, Long Island, New York reads "*Abies canadensis atrovirens* (deep green-leaved hemlock) A dwarf form with conspicuously small foliage as dark as that of the Yew".

  This might well have been the Beissner plant, but a tree at the Arnold Arboretum received from the Parsons Nursery in November, 1880, although grown there under this name, clearly cannot be this clone, since it bears leaves that are remarkably short and wide, but not particularly dark green. An acceptable name should be registered if this tree is to be propagated.

  The name 'Atrovirens' has also been widely but incorrectly applied to a cultivar with leaves tinged yellow-green, distributed by the Hicks Nursery, Westbury, New York. For this a new name has had to be found. See 'Hicks'.

  I have otherwise been unable to trace the true 'Atrovirens' which must be regarded as probably lost to cultivation, at any rate in America. So any fresh selection within the description should be registered under a new name.
- - f. **'aurea'** See p. 73.
- - **'Aurea'** Nelson, Pinaceae 32. 1866. (*Abies canadensis aurea*). GOLDEN GROUP. "The golden leaved" form. Since the clone originally bearing this name cannot now be identified, it should not be used in any clonal sense. Plants in collections thus labelled are usually 'Everitt Golden'.

T. canadensis 'Beaujean'.

T. canadensis 'Bacon Cristate'. Home of Ralph Bacon, Sparta, New Jersey.

T. canadensis 'Bennett'. Garden of Ralph Lott, Eatontown, New Jersey, in 1938. The original plant. J.C.S. No. 1376.

- - 'Aurea Compacta' Hort. Indistinguishable from 'Everitt Golden'.
- - 'Aurescens' Hort. See 'Silvery Gold'.
- - **'Bacon Cristate'** Registered name. Smith ex Hebb, Arnoldia. **23:** 208, 1973. DENSE LEAF GROUP. This is one of a number of extremely dwarf cristate hemlocks found in the wild in northern N.J. about 1925 by the late Ralph Bacon. It resembles 'Jervis' but it is much more cristate and dwarf and the foliage is dark green. Formerly known as Bacon No. 1, it was registered and introduced by Don Smith (Watnong Nurs. Cat. 1966).
- - **'Bagatelle'** New cultivar. GLOBOSE GROUP. A multiple-stem plant that grows rapidly. According to Joseph Cesarini, it develops into a compact globose plant more than 3 feet high in four years from grafts, without shearing. This cultivar originated in the Bagatelle Nursery, Center Moriches, Long Island, New York.
- - **'Baldwin Dwarf Pyramid'** New cultivar. Listed Spingarn, Nurs. Cat. 1970 (Name only) CONICAL GROUP. This I have only seen as an upright, rather open bush less than 1 m high. The foliage is noticeably variable in size, the leaves ranging from 15 mm $\times$ 2 mm long and widely spaced on the older shoots, graduating towards the growth tips where they become very much smaller and radially held, the ultimate truss of tiny leaves hiding the bud. The growth is dense but not sufficiently so to hide the shoots which are light-brown and minutely pubescent.
- - 'Baldwin Dwarf Globe'. Provisional name. Listed Spingarn, Nurs. Cat. 1967 (Name only.)
- - **'Barrie Bergman'** New cultivar. SPREADING GROUP. A flat-topped, slow-growing variant with foliage similar to 'Bennett' but with slower growth and shorter leaves. Propagated by the late Fred Bergman, Feasterville, Pennsylvania, from a bud sport on 'Cole' about 1970, and named after his daughter. In 1973 it was about 12 inches high and 14 inches wide.
- - 'Baumann'. Provisional name. Listed by Kluis c. 1970 (Name only.)
- - **'Beaujean'** Hillier, Dwarf Conif. 66. 1964; 'Saratoga Broom' Swartley ex den Ouden and Boom, Man. Conif. 462. 1965. SPREADING GROUP. The branches spread symmetrically upward and outward, forming a cone-shaped cavity in the center, with lateral branchlets much shorter than the terminal, more pubescent and darker reddish-brown twigs and shorter leaves (3-9 mm), a lighter green color and less persistent (1½ years). Found by A. C. Beaujean of Yonkers Nursery, Yonkers, New York, in 1926, as a witches' broom on a lateral branch of a small hemlock in a wood near Saratoga, New York. In 1928 he sent scions to the Newport Nurseries, Newport, Rhode Island, later transplanting the entire tree to the Yonkers Nursery, where the broom eventually became so heavy that the branch split and died. This cultivar is commonly found in collections with different spellings of the name.
- - **'Beehive'** New cultivar. GLOBOSE GROUP. This forms a dense, bee-hive-shaped bush with fine-textured foliage with small green leaves and at least a tendency towards the 'Cinnamomea' type of colouring. The origin is not recorded, but it was first listed in the Mitsch Nursery Cat. 13. 1982.
- - **'Bennett'** Swartley ex den Ouden and Boom, Man. Conif. 450. 1965; 'Bennett's Minima', Hillier, Dwarf Conif. 66. 1964. SPREADING GROUP. A low, spreading, compact shrub much broader than high, with branches more or less arching over or horizontal with pendulous tips; lateral branchlets about as long as the terminal. Twigs are slender and borne in flat sprays, creating a fanlike effect. Final side shoots all held at 60° to the main shoot. Leaves are fairly densely held, to 12 mm $\times$ 1.75 mm on old shoots, less on young growth, elliptic, ending in a broad sharp tip. One unusual feature is that the leaves on the top of and lying reversed along the shoot are in many cases as large as the other leaves on that shoot.

　　Mr. Bennett, a nurseryman of Atlantic Highlands, New Jersey, imported a few "Japanese" hemlocks about 1920, reputedly from Japan. One of these plants turned out to be a distinguished, spreading Canada hemlock, which was later acquired by Ralph Lott of Eatontown, New Jersey. It was propagated from grafts and cuttings by W. H.

Wittenberg, Long Branch, New Jersey who, regrettably, caused confusion by distributing the plant as 'Minima'. The grafted plants were comparatively taller than those rooted from cuttings, and the foliage was noticeably coarser.

This is one of a group of plants arising in seed-beds from time to time that are virtually indistinguishable. See also 'Minima' below. The European 'Minima' from Hesse, according to Mr. Hohman, is not the same as the 'Minima' of either Warner, Gotelli, Bennett, or Wittenberg. Mr. Hohman claimed his 'Minima' from Hesse is different from and better than 'Bennett'.

- - 'Bennett's Minima' Hort. See 'Bennett'.

The following group of cultivar names exemplify the problems discussed on p. 84 created by the coining of names consisting of two words, the first recording the origin or the originator and the second indicating a group classification. Since these names have never been "validly published", in a technical sense, I take the opportunity of ratifying here some names of acceptable form that have come into use or naturally suggest themselves. The remainder are listed as "Provisional". This leaves the way open for such clones as may still prove to be in cultivation to be registered under more satisfactory names.

- - 'Bergman's Aurescens Nana' Listed Raraflora Nurs. Cat. c. 1970. (Name only.) See 'Silverygold'.
- - 'Bergman's Cinnamon Dwarf' Provisional name. Listed Raraflora Nurs. Cat. c. 1970 (Name only.) CINNAMON-TIP GROUP. The standing and value of this and some of the following names is very uncertain. They are selections made by the late Fred W. Bergman and appear as names only in later catalogues of Raraflora Nursery, Feasterville, Pa. The collection at Raraflora was dispersed at an auction sale after his death and I have no information relating to the purchasers. It is probable that some names relate to a plant that had not been propagated and which may have been sold unlabelled, in which case they will never be identified. If this clone can still be identified, a name acceptable for registration should be found.
- - 'Bergman's Contorta Spicata' Raraflora Nurs. Cat. c. 1970. See 'Contorted'.
- - 'Bergman's Dwarf Long-Leaf'. Provisional name. Listed Raraflora Nurs. Cat. c. 1970 (Name only.) A dense-leaf dwarf with growth but a little over 1 inch per year; somewhat irregular. It is characterised by small whorls of tiny leaves near the growing tips. A small plant at Raraflora, Feasterville, Pennsylvania was 13 inches high and 15 inches across in 1970. Three years later it was 20 inches high and 21 inches across. See above, under Bergman's Cinnamon Dwarf. If this clone can still be identified, a name acceptable for registration should be found.
- - 'Bergman's Dwarf' Provisional name. Listed Raraflora Nurs. Cat. c. 1970 (Name only.) See above, under 'Bergman's Cinnamon Dwarf'. If this clone can still be identified, a name acceptable for registration must be found.
- - 'Bergman's Dwarf Upright' See 'Bergman's Dwarf'.
- - 'Bergman's Fantana Broom' Raraflora Nurs. Cat. c. 1970. See 'Fantana Broom'.
- - 'Bergman's Frosty' Raraflora Nurs. Cat. c. 1970. See 'Frosty'.
- - **'Bergman's Gem'** New Cultivar. Listed Raraflora Nurs. Cat. c. 1970 (Name only.) CINNAMON-TIP GROUP. This forms a globose plant, similar to but more compact than 'Cinnamomea'. It originated with a plant found "in the wild" in 1964 in Vermont. A grafted plant was 8 inches by 10 inches in 1970 and three years later was 14 inches by 14 inches.
- - 'Bergman's Gold' Hort. See 'Silverygold'.
- - 'Bergman's Heli' Raraflora Nurs. Cat. c. 1970. See 'Heli'.
- - 'Bergman's Hussi-sib' No longer identifiable or listed.
- - 'Bergman's Spreading' Provisional name. Listed Raraflora Nurs. Cat. c. 1970 (Name only.) Similar to 'Cole's Prostrate', but stronger growing. If this clone can still be identified, a name acceptable for registration should be found.

- - 'Bergman's Wee-leaf' Provisional name. A plant of moderate growth that originated in a bud-sport in a hedge. The leaves are 3-6 mm long. See above, under 'Bergman's Cinnamon Dwarf'. If this clone is still identifiable, a name acceptable for registration should be found.

- - 'Bergman's Wee-tip' Provisional name. Raraflora Nurs. Cat. c. 1970. (Name only.) A good irregular semi-dwarf, dark green seedling collected by Mr. Bergman. The leaves, two-ranked above and scattered below, are short near the tips of the lateral twigs, suggesting the name. Shaded leaves tend to be glossy and average wider and shorter than typical hemlock. In 1970 the original plant was 4 feet high and 5 feet wide. Three years later it measured 4 feet 10 inches high and a little over 6 feet wide.

    If this clone can still be identified, a new name acceptable for registration should be found.

- - **'Betty Rose'** New cultivar. WHITE-TIP GROUP. Very compact, very irregular bush, annual growth 2-3 cm; branchlets very pubescent near tips; leaves very irregularly disposed and pointing forward more than usual, 4-14 mm long, with prominent teeth, narrow, the longer leaves little more than 1 mm wide; lines of stomata rather narrow.

    Spring growth, which is the primary growth, and late summer growth, which is the secondary growth, are clear white (All other known clones are white on the secondary growth, not the primary, thereafter gradually fading in early fall.)

    The original plant was discovered, when quite small, in Vanceboro, Maine, by Francis J. Heckman of Ambler, Pennsylvania. In 1973 it was about 61 cm high and 91 cm wide. This plant has been propagated and as of 1983 has not grown to any extent. Since there is a seven year difference in the two times Mr. Heckman visited the area in Maine the date of its collection is uncertain by that amount of time, a growth description must be taken from a cutting rooted in 1967. This cutting has grown to a height of 15 cm and a width of 30 cm during the past fifteen years. 'Betty Rose' is the dwarfest of the white-tipped hemlocks, a very striking plant and a favorite of the author.

- - 'Biltmore Weeping' Provisional name. Listed Raraflora Nurs. Cat. c. 1970 (Name only.) See above, under 'Bergman's Cinnamon Dwarf'.

- - **'Bonnie Bergman'** New cultivar. Listed Raraflora Nurs. Cat. c. 1970 (Name only.) SPREADING GROUP. Propagated from a terminal spreading witches' broom on a small seedling 30 inches high. It is bare in the middle with twisted, sculptured bark. Comparatively slow-growing and difficult to propagate. The plant was obtained from a Christmas tree grown near Stroudsburg, Pennsylvania, in the autumn of 1972 by the late Fred Bergman, Pennsylvania, and is named for one of his daughters.

- - **'Boulevard'** Swartley ex den Ouden and Boom, Man. Conif. 451. 1965. CONICAL GROUP. Pyramidal, very compact; branches irregular in length; inner branchlets drooping; leaves somewhat densely set and longer than normal hemlock; very dark green. It originated as a selected seedling in the Boulevard Nurseries, Newport, Rhode Island, in about 1930. Den Ouden and Boom stated that it is no longer in cultivation, but this is incorrect. The original plant is at Far Country. In 1938 it was nearly 2 feet high and 1 foot 6 inches wide. In 1950 it was about 10 feet high and 6 feet 6 in width. At Raraflora, Feasterville, Pennsylvania, a fifteen-year-old plant was over 6 feet high. A tree at U.S. National Arboretum (Accession No. 24578) is 6 feet high $\times$ 5 feet wide, is so compact and dense as to appear to have been sheared. 'Boulevard', because of its dark green colour and graceful habit, is a striking plant.

- - 'Bowman' Provisional name. I understand that this name was applied to one of Joseph F. Gable's selections, but I have been unable to trace it in current cultivation.

- - 'Bradley' Hort. See 'Brandley'.

- - **'Bradshaw'** Hohman ex Swartley in Amer. Nur. **133**(8): 12. 1946. CONICAL GROUP. A good compact pyramid with some irregularity in branching and leaf arrangement, branches spreading with tips nearly upright; leaves bright green. Selected from nursery seed-bed by Henry J. Hohman of Kingsville, Maryland. It makes a beautiful specimen and is recommended for hedging.

T. canadensis 'Bradshaw'. Kingsville Nurseries, Kingsville, Maryland, in 1945. The original plant. J.C.S. No. 1820.

T. canadensis 'Betty Rose'.

T. canadensis 'Brandley'. Gotelli collection, U.S. National Arboretum, Washington, D.C. Accession No. NA 21106.

T. canadensis 'Bristol'. Briston Nurseries, Bristol, Connecticut, in 1945.

T. canadensis 'Callicoon'. Longwood Gardens, Pennsylvania.

- - **'Brandley'** Dawson ex Jenkins, Heml. Arb. Bull. 15. 1936. CONICAL GROUP. A compact, broad, conical form with a tendency for the leader to divide. The original was a seedling selected by Mr. Brandley of Walpole, Massachusetts, about 1920. Later it was distributed by the Eastern Nurseries, Holliston, Massachusetts. In 1938 a plant at Far Country was 18 inches high and 18 inches wide. Seven years later it was 4 feet 3 inches high and a little less in breadth.

  Two plants at the U.S. National Arboretum, Washington, D.C., (Accession Nos. 21106 & 21107) are respectively 6 feet high × 6 feet and 10 feet high × 6 feet wide. Globose to conical habit, dense uniform growth. Needles medium to dark green.
- - 'Brandley Broom' Hort. See 'Narragansett'.
- - **'Brevifolia'** Headfort in Chittenden, Report Conif. Confce. 546. 1932. LITTLE-LEAF GROUP. According to information received by Mr. Jenkins in a letter from the Marquess of Headfort, Kells, County Meath, Ireland, the leaves are no longer than those of 'Parvifolia', 3-8 mm, but the branchlets are heavier. 'Brevifolia' was purchased by the Marquess of Headfort from James Veitch and Sons., Chelsea, London, in 1913. At seventeen years it was 9 feet high and 9 feet wide.
- - **'Brimfield Golden'** Provisional name. GOLDEN GROUP. A fast-growing, sparsely branched, yellowish hemlock from Brimfield Gardens Nursery, Wethersfield, Connecticut. It would hardly seem sufficiently distinctive to justify retention of the name.
- - **'Bristol'** Swartley ex den Ouden and Boom, Man. Conif. 451. 1965. GLOBOSE GROUP. Compact, broad-topped, bushy shrub; branches horizontal with good light-and-shadow effect; branchlets broadly fan shaped, leaves very blunt. This originated in the Bristol Nurseries, Bristol, Connecticut. Propagation has been limited. There is a good plant in the collection of Swarthmore College, Swarthmore, Pennsylvania (Accession No. 55-23).
- - 'Bristol Short-Leaf' Provisional name. A free-growing plant similar to 'Jenkinsii'. One at Raraflora, Feasterville, Pennsylvania, was 11 feet 6 inches high in 1970 (from Bristol Nurseries, Bristol, Connecticut). According to Alex Cumming, Superintendent of the Bristol Nurseries, in a letter of September 28 1938, they obtained propagating material of both LITTLE-LEAF and LARGE-LEAF clones from an estate near Stamford, Connecticut.

  On the above information it is not clear that this clone is sufficiently distinctive to justify registration. If it is, a new name should be found that will avoid confusion with the cultivar 'Bristol'.
- - **'Brookline'** Welch, Dwarf Conif. 323. 1966. PENDULA GROUP. This is one of the original seedlings which General Howard is said to have discovered in the mountains back of Fishkill, New York. The original 'Brookline' hemlock at Holm Lea, the home of Charles S. Sargent, is now on the property of Mrs. Roger Ernst, facing Sargent Drive in Brookline, Massachusetts, about two miles from the Arnold Arboretum. The tree has a horizontal system of branching and is moundlike, with the new growth following the contour of the branches.
- - **'Broughton'** Swartley ex den Ouden and Boom, Man. Conif. 451. 1965. LARGE-LEAF GROUP. A slow-growing pyramidal plant of irregular habit; leaves rather long, densely set. A plant at Far Country, a graft of 1933, was 1 foot 3 inches high and 1 foot wide in 1938. In 1970 it was 13 feet high and 10 feet 6 inches wide. Its character is now rather loose. 'Broughton' originated in the Nursery of Leslie M. Broughton, Madison, Ohio. 'Brandley', a selection of the same habit but with shorter leaves, is a better plant.
- - **'Callicoon'** Swartley ex den Ouden and Boom, Man. Conif. 451. 1965; var. *callicoonii* Curtis ex Jenkins, Heml. Arb. Bull. 17. 1937. WEEPING GROUP. This attractive cultivar arose as a seedling on the Curtis Nurseries at Callicoon, New York. It is a moderately vigorous shrub nearly twice as broad as high. The branches stand out at angles varying between horizontal at the sides of the plant to 60° upwards at the centre, with all the terminal shoots pendulous, resulting in a habit of growth reminiscent of the well-known *Juniperus media* 'Pfitzerana'. It has been widely, but incorrectly, distributed in the trade as 'Curtis Spreader'.

- - 'Calvert' A listed name. I have no further information.
- - **'Cappy's Choice'** Mitsch, Nurs. Cat. 12. 1974. PENDULA GROUP. A compact low-growing clone with fine-textured foliage; light green with a hint of gold. A seedling selected by James Caperci, Rainier Mount Alpine Gardens, Seattle, Washington 98168. Will reach 18 inches in ten years.
- - 'Cesarini's Broom' Provisional name. This is represented only by a small plant in the Arnold Arboretum, Accession No. 1404-67. A plant grafted by Joseph Cesarini closely resembles 'Cole's Prostrate'. No doubt time will indicate whether this merits recognition as a new cultivar.
- - 'Christie' Hort. See 'Muttontown'.
- - **'Cinnamomea'** Hort. ex Jenkins, Heml. Arb. Bull. 6. 1934 ("cinnamomeus"); Abbott ex Swartley, Amer. Nurs. **133**(7): 7. 1946. CINNAMON-TIP GROUP. Leaves near the tips of the sprays are pointed and whorled; mature leaves are blunt. Two identical plants were found by Frank L. Abbott near Athens, Vermont, in 1929. A plant at Far Country was 21 inches high and 15 inches broad in 1938. In 1945 it was 31 inches high and 48 inches in breadth. This cultivar has been widely propagated.
- - **'Cloud Prune'** Registered name. Smith ex Hebb, Arnoldia **33:** 209. 1973. SPREADING GROUP. The original plant was discovered by William D. Wallbridge, years ago, it was the size of a basketball and very dense. Until eight years ago it grew very slowly, then began to grow much more rapidly and is now an irregular spreader. The leaves are quite irregularly disposed. In 1972 the original plant was about 3 feet high and 5 feet wide. It has developed in a fashion suggestive of Japanese "cloud pruning", hence the choice of name.
- - **'Coffin'** New cultivar. TWIGGY GROUP. Similar to 'Jervis', but the branches are thinner and the leaves less crowded. The congested branchlets terminate in twigs that have few if any laterals. This plant has been propagated and distributed by Eastern Nurseries, Holliston, Massachusetts. A plant in the Arnold Arboretum, Accession No. 129-72, was 1 foot high in 1973.
- - **'Cole's Prostrate'** Gotelli, Wm. T., Amer. Hort. Mag. **39**(4):187. 1960; 'Cole' Swartley, Plants and Gar. **21:** 17. 1965; Welch, Dwarf Conif. 319. 1966. 'Cole's Prostrate', Helene Bergman, Plants and Gar. **21:** 43. 1965. SPREADING GROUP. This is the smallest of the weepers, a true rock-garden jewel. The original plant was collected by H. R. Cole in 1929 near the bottom of Mt. Madison, Coos County, New Hampshire. It was transplanted to the nursery of Gray and Cole near Haverhill, Massachusetts, where it was observed by the author in 1938. It was then slightly less than 6 inches in height and 24 inches in breadth. The leaves are closely set, pectinate, reducing in length towards the tip of the shoot, uniformly held at a wide angle to the shoot, with two noticeably white stomatic bands. The main branches are horizontal and are devoid of leaves and the branching habit is most peculiar: the branchlets grow outward but soon knuckle under and bend directly downward, so most of the tips are in close contact with the soil. These tips have a tendency to rot. A mulch of pebbles is of some help, but the plant grows more rapidly and is perhaps more beautiful if staked to a height of about 12 inches and then allowed to weep downward like a vernal fountain. This clone is difficult to establish unless given some shade. Two thirty-year-old plants were at one time at Raraflora, Feasterville, Pennsylvania. One, unstaked was 4 inches high and 25 inches in diameter. The other, which had been staked for a time, was 33 inches high and 25 inches across. Confusion between this little gem and the strong-growing plant mis-named 'Prostrata' (p. 130) is ridiculous.
- - **'Columnaris'** Bolle ex Beissn. Handbuch Conif. 402. 1891. FASTIGIATE GROUP. A columnar cultivar found in Germany many years ago by Dr. Bolle. This apparently is no longer in cultivation. See also 'Harmon'.
- - **'Compacta'** Sénéclauze, Conif. 19. 1868. Named from a plant that arose in the nursery at Bourg-Argental, Loire, France, which formed a conical pyramid with slender branches and scattered, small leaves (3-5 mm long) which are a beautiful green, glaucous on the

T. canadensis 'Cinnamomea'. Far Country, Philadelphia, Pennsylvania, in 1945. J.C.S. No. 1348.

T. canadensis 'Cole'.

underside. It is probably no longer in existence. Many different clones of hemlock have been distributed under the name 'Compacta' with much resulting confusion. For example, Curtis Nurseries, Callicoon, New York, at one time sold several different clones of Canada hemlock under the name 'Compacta' and Hillier (Dwarf Conif. 66. 1964) refers to yet another clone in the Gotelli collection under this name. All this is further evidence of the need to assign fancy cultivar names to superior variants of hemlock and the need to register new names (See 'Pomfret' and 'Stockman Dwarf').

- - 'Compacta Aurea' Hort. See 'Everitt Golden'.

- - **'Compacta Nana'** Beissn. Handbuch Nadel. 402. 1891. Another early European selection, described as a dwarf, compact, slow-growing shrub; densely branched, leaves small. This is no longer in cultivation.

- - **'Conica Nana'** Young ex Hornibrook, Dwarf Conif. ed. 2, 272. 1938; *Abies canadensis nana* Young, Nurs. Cat. 6. 1875. A dwarf, conical form with drooping shoot ends that originated in the nursery of M. Young, Guildford, Surrey, England. This, also, is no longer in cultivation.

- - 'Connecticut Turnpike' Provisional name. TWIGGY GROUP. This is a dense-leaf, irregular dwarf with a crested appearance propagated from an unidentified plant discovered by Greg Williams and Layne Ziegenfuss along the Connecticut Turnpike near exit 55, in the collection of William Patton, Downingtown, Pennsylvania. It can only be named provisionally in case it later turns out to be an already named clone.

- - 'Contorta Spicata' Hort. See 'Contorted'.

- - **'Contorted'** Helene Bergman, Plants and Gar. **21:** 44. 1965; 'Bergman's Contorta Spicata' Raraflora Nurs. Cat. (prior to 1959). Upright, bushy habit with an open center; branchlets of the current growth twisted downwards and clustered at the tips of the previous year's growth; characterized by the twisted tips of many twigs; leaves dark green, broader than on typical hemlock. Selected as a seedling in Carroll's Nursery, Neshaminy, Pennsylvania, about 1959. A five-year-old plant is 18 inches high.

- - 'Copland' Hort. See 'Coplen'.

- - **'Coplen'** Swartley ex den Ouden and Boom, Man. Conif. 453. 1965. CONICAL GROUP. A compact, pyramidal plant, growing at about half the rate of typical hemlock. Branches and branchlets somewhat more crowded; leaves rather broad. Found in 1921 by M. G. Coplen, Rockville, Maryland, in the vicinity of Mount Storm, West Virginia. Propagation has been limited. This cultivar grows much better in wet, heavy soil than typical hemlock, as can be observed in the U.S. National Arboretum, and it is not lacking in grace. A plant at Raraflora, Feasterville, Pennsylvania, about twenty-five years old, was 8 feet high and 7 feet wide in 1972. It does not retain the shape of a compact pyramid unless it is severely trimmed.

- - **'Corbit'** New cultivar. Listed Hillside Nurs. Cat. c. 1970 (Name only.) SPREADING GROUP. Similar to 'Bennett' in habit but slower growing and denser. Obtained by Joseph Cesarini of Sayville, Long Island, from Joseph Popeleski, Greenlawn, New York. In 1970 the original plant was 15 inches high and 25 inches wide. Named in honour of the late Dr. John D. Corbit, Penn Valley, Pennsylvania.

- - **'Creamey'** New cultivar. WHITE-TIPPED GROUP. This is another plant originating with Otto Gentsch. It is very similar to 'Gentsch White' save that it is more compact and has the tip growth cream colour, not white.

- - **'Curly'** Epstein ex Spingarn, Amer. Rock Gar. Soc. Bull. 27: 86. 1969. This interesting clone forms an upright and compact bush with most unusual leaves. These are short and relatively broad (7 mm by 2 mm) rounded at the tip, held pectinately at 90° from the shoot, more or less in two ranks, dull green above and with two stomatic bands beneath lying close to the midrib with wide green margins giving the plant a greyish cast. The peculiarity is that the twisting of the short leaves on top of the shoot that is characteristic of the species is quite absent. The name is suggested by the tendency of each leaf to curl downwards, giving a very attractive appearance. Harold Epstein of Larchmont, New York, found the original in the garden of a friend, took cuttings, and raised a limited number of plants. The foliage is similar to 'Verkades Recurved', but is stronger growing. The branches in both clones are brittle. A plant at the National Arboretum, Washington, D.C. (Accession No. 36294) is 3½ feet high and 3½ feet wide.

The following group presents considerable difficulties.

Firstly, these names exemplify the problems discussed on p. 84 created by the coining of names consisting of two words, the first recording the origin or the originator, and the second indicative of a group classification.

Secondly, they all arose as selections from a large batch of interesting plants resulting from seed sown at the Curtis Nursery, Callicoon, New York, in 1926. Unfortunately, many of these seedlings were distributed before the question of names arose, so the identification of any names with particular clones is very difficult. Two cases that are clear of both of these objections are 'Callicoon' and 'Curtis Ideal'. These are here retained but the remainder are listed as 'Provisional' names only. This leaves the way open for the later selection of particular clones and the designation of particular trees as clonotypes and their registration under more acceptable names.

- - 'Curtis' Curtis ex Jenkins, Heml. Arb. Bull. 17. 1937. (as var. *curtisii*) A broadly conical plant in sharp outline. I have been unable to identify this clone.
- - 'Curtis Compact' Provisional name. A seedling raised in the Curtis Nursery, Callicoon, New York, about 1926. In 1983 it is 10 feet high and 6 feet wide. The leaves are ⅜ inches long and ⅛ inches wide, very dark green, densely set in double rank. The plant has a fairly smooth appearance. A more acceptable name should be registered.
- - 'Curtis Dense-Leaf' Provisional name. This is another of the seedlings. This is a vigorous clone, growing 10-15 cm per year. The mother plant is now a plant 24 feet high and 18 feet wide at Swarthmore College, Swarthmore, Pennsylvania, Accession No. 9357. It will tolerate full sun and maintains a good compact habit without pruning. Many Dense-Leaf seedlings originated in the Curtis Nurseries, Callicoon, New York, some regrettably having been distributed under the name 'Compacta'. A more acceptable name should be registered for any clone considered worthy of it.
- - **'Curtis Ideal'** Swartley ex den Ouden and Boom, Man. Conif. 453. 1965. CONICAL GROUP. Dwarf, conical, irregularly shaped, broader than high. Leaves bright green as long as in 'Macrophylla', not quite in two ranks, not closely set. This is a remarkable cultivar and one easily recognised because of the unusual combination of characters. In 1946 two plants that had been grafted in 1935 averaged 30 inches in both height and breadth.
- - 'Curtis Largeleaf' Provisional name. Maintains an upright compact habit even when mature. Represented at Swarthmore College, Swarthmore, Pennsylvania, Accession No. 1286, near the walk entering the tunnel under the railroad tracks. A name more acceptable should be registered if this clone is to be brought into circulation.
- - 'Curtis Little-Leaf' Provisional name. Selected from seed sown in 1926 at the Curtis Nurseries, Callicoon, New York. In 1936 the mother was about 8 feet high. A better name should be found if this selection can be identified and justifies registration.
- - 'Curtis Sparse-Leaf' Provisional name. This name has been provisionally assigned to a good, compact plant at Far Country, at least forty years old, which is now 6 feet high and about 8 feet wide. There are about 9 leaves per centimeter of branchlet. It is irregular in habit because of variability in length of twig growth, and it remains compact without pruning. A better name should be registered if this selection justifies registration.
- - 'Curtis Spreader' Hort. and 'Curtis Low Spreading' Hort. See 'Callicoon'.
- - 'Curtis Weeping' Provisional name. An upright plant with the topmost branches markedly pendulous. The leaves on the tips of the branchlets are sparse. Grown from seed sown in 1926; about 9 feet high in 1938, nearly 35 feet high in 1970. A better name should be found if this selection justifies registration.

- - 'Curtis Wide-Leaf' Provisional name. The plant at Far Country should be considered the clonotype of this variant. Large plants develop an exotic appearance, due to both habit and foliage, the branchlets having a tendency to lie in various planes and the leaves being very obtuse or notched at the apex and very broad, 2-3 mm wide, and being somewhat shorter than typical hemlock. About seven trees of the same character were observed in the Curtis Nurseries, Callicoon, New York, in 1945 with two trunks, from seed sowed in 1926. The foliage is almost identical to that of 'Wilton Wideleaf'. A better name should be registered if this selection justifies retention.

- - 'Cushion' New cultivar. CINNAMOMEA GROUP. This was named for a bun-shaped plant with a neat appearance (as though sheared) of unknown origin in the U.S. National Arboretum at Washington D.C. (Accession No. 20859) 2½ feet tall × 3½ feet across. The foliage is similar to 'Rugg's Washington Dwarf', from which it is doubtfully distinct enough to be needed in any but a specialist collection. The needles are densely held, stiff, ½ inch long, down to ¼ inch long at the branch tips.

- - 'David Verkade' New cultivar. PENDULA GROUP. The appearance of this clone is between 'Cole' and 'Sargentii'. The original plant has recently died, but it had been propagated by the Verkade Nursery, Wayne, New Jersey.

- - 'Dawsoniana' Hatfield ex Bailey, Cult. Conif. 124. 1933. WIDE-LEAF GROUP. This clone forms a compact, upright tree with rather crowded branches regularly ramifying in flat sprays. The leaves are up to 10 mm long, wide (to 1.5 mm), parallel-sided with rounded ends, two-ranked, arranged pectinately and held uniformly at 80° to 90° to the shoot; 2 broad bands of stomata; a bright, medium green. The minute buds are invisible. This clone was propagated by grafting from a compact plant in the Hunnewell Arboretum about 1926, by the owner of Eastern Nurseries, Henry S. Dawson. The plant was observed there in 1938, measuring 3 feet 8 inches in height and 5 feet in breadth. As it becomes older it grows more upright. It is not greatly different from typical hemlock but can always be identified in collections because of its attractive and distinctive foliage. It has the most attractive foliage of any clone in this group.

- - f. densifolia. See p. 73.

- - 'Densifolia' Jenkins, Heml. Arb. Bull. 6. 1934. Swartley ex den Ouden and Boom, Man. Conif. 454. 1965. This name has been loosely applied to numerous plants, so it should not be used in any clonal sense.

- - 'Detmer's Weeper' Don Smith, Watnong Nurs. Cat. 11. 1972. PENDULA GROUP. This is supposedly a seedling of Sargent hemlock. The name was suggested by Joseph Cesarini, who obtained scions from a plant 18 inches high and 12 feet wide at Detmer's Nursery, Tarrytown, New York. The original plant has since died, but Mr. Cesarini had a grafted plant 12 inches high and nearly 5 feet wide. It appears to be intermediate between 'Sargent' and 'Cole', showing some bark on the top as in 'Cole'.

- - 'Ditmar's Weeper' Hort. See 'Detmer's Weeper'.

- - 'Doc's Choice' New cultivar. Reg. No. 1010 CONICAL GROUP. A clone selected by Robert F. Doren, U.S. National Arboretum, Washington D.C. It forms a slow-growing, neat, conical plant with tight, dense foliage of a much darker green than normal. It reached 4.5 m in 20 years.

- - 'Doran' Helene Bergman, Plants and Gardens 21: 42. 1965. CONICAL GROUP. Forms a compact and symmetrical cone with a sheared effect when young; branches crowded and slightly pendulous, ascending tips; leaves nearly normal. The mother plant was discovered by Professor William Doran of the University of Massachusetts, in a cow pasture in Vermont. A propagation reached 30 inches high and 32 inches at fifteen years, but later grew more vigorously. The foliage is light green; the needles very densely arranged on the branchlets.

- - 'Doran's Dwarf' Probably a mistaken name for 'Doran'.

- - 'Doran's Prostrate' Provisional name. Only one plant has been observed by me, in the U.S. National Arboretum, Washington D.C. (Accession No. 21780). In 1970 it was 3 feet 10 inches high and 9 feet wide. It was a decorative plant very similar to 'Curtis

T. canadensis 'Coplen'. U.S. National Arboretum, Washington, D.C. in 1970. J.C.S. No. 1418.

T. canadensis 'Corbit'. Collection of Joseph Cesarini, Sayville, Long Island, New York, in 1970. J.C.S. No. 1833.

T. canadensis 'Curly'.

T. canadensis 'Curtis Dense-Leaf'. Swarthmore College, Swarthmore, Pennsylvania, in 1972. Accession No. 9357.

T. canadensis 'Curtis Ideal'. U.S. National Arboretum, Washington, D.C.

T. canadensis 'Curtis Large-Leaf'. Swarthmore College, Swarthmore, Pennsylvania, in 1972. Accession No. 1286.

T. canadensis 'Curtis Sparse-Leaf'. Far Country, Philadelphia, Pennsylvania, in 1975. J.C.S. No. 1847. Photo by F. Ray.

Spreader'. The plant had been received from Ralph M. Warner of Milford, Connecticut, a discovery in the wild about 1958 near Amherst, Massachusetts, by Professor William Doran of the Department of Botany of the University of Massachusetts. If considered sufficiently distinctive for registration, a name should be found that does not incorporate the existing cultivar name 'Doran', since that would be almost certain to cause confusion.

- - 'Dr. John Corbit' See 'Corbit.'
- - 'Dr. Peter Hornbeck' See 'Hornbeck.'
- - 'Dr. John Swartley See 'John Swartley'.

The *Cultivated Code* Recommendation 31A strongly disfavours names containing abbreviations and forms of address.

- - **'Dover'** New cultivar. LARGE-LEAF GROUP, Dirk van Heiningen, proprietor of the van Heiningen Nurseries, Dover, Pennsylvania, collected seed from the South Wilton Nursery specimen of 'Macrophylla' about eighteen years ago and grew several hundred seedlings. These are quite variable in habit of growth and foliage. A compact plant with large leaves, 4 feet 8 inches high and the same in breadth, was selected as clonotype.
- - **'Drake** Palette Gardens Nurs. Gar. 1978 (As 'Drake Spreader') CINNAMOMEA GROUP. A clone of the cinnamomea type which makes a low bush with a central depression. Cinnamon-brown hairs on young twigs. The simpler name is preferable.
- - 'Dwarf Upright' Hort. See 'Lewis'.
- - 'Dwarf Pyramid' See 'Baldwin Dwarf Pyramid'.
- - **'Dwarf White-tip'** Swartley ex den Ouden and Boom, Man. Conif. 454. 1965. WHITE-TIP GROUP. This is rather a misnomer, because this clone develops into a graceful, broad, conical, but by no means dwarf shrub of quite open habit, with leaves white on summer growth, tending to fade in autumn. The foliage is in flat sprays, the leaves (to 15 mm × 1.75 mm) being densely borne and carried pectinately, each leaf standing away from the shoot at 60-70° (quite uniformly throughout the bush). The original plant is in the Morris Arboretum. In 1938 it was 7 feet high and 10 feet wide. At the present time, part of the plant is growing faster and thus losing some intensity in the whitish leaves. It probably came from the Arnold Arboretum about 1890, although there is no record of such a plant at that institution. 'Dwarf White-tip' is a very interesting and decorative plant. It has been widely propagated. It was at one time being distributed by Kingsville Nurseries, Kingsville, Maryland, as 'Albo-spica'.
- - 'Ehrle Large-Leaf' Provisional name. This clone has long leaves that are densely set. The Arnold Arboretum received this plant from Millcreek Nursery, Newark, Delaware, Accession No. 788-58. It had been propagated from a tree at Swarthmore College, Swarthmore, Pennsylvania, which has since disappeared. It presently has a compact, slender habit and in 1970 was 6 feet high and 2 feet 6 inches wide. The foliage is dark green. The narrow habit may not persist.
- - 'Elciclis' A listed name not worth retaining.
- - **'Elm City'** Swartley ex den Ouden and Boom, Man. Conif. 454. 1965. PENDULA GROUP. A semi-weeping, non-symmetrical bush with several main branches growing upward and outward. It is usually broader than high, but grafted plants of the same age may vary considerably in height. The original plant at Far Country was received in 1932 from the Elm City Nursery, Elm City, Connecticut. It was then 2 feet high. Fifteen years later it was 9 feet high and 11 feet wide. In 1970 it was 15 feet high and 22 feet wide. Cones are sometimes produced in clusters. It is an interesting, picturesque plant.
- - **'Essex'** New cultivar. CONICAL GROUP. The original plant was discovered by Gregory Williams at Essex, Vermont. It has foliage like 'Pygmaea' but with somewhat smaller leaves. It forms an irregular upright pyramid. The leaves are light green, annual growth ½ to ¾ inches. According to the records of the Arnold Arboretum, the original plant was less than 3 feet high at the age of about forty years.

- - 'Everitt Dense-Leaf' Hort. Indistinguishable from 'Hussii'.

- - **'Everitt Golden'** Swartley ex den Ouden and Boom, Man. Conif. 454. 1965. GOLDEN GROUP. Stiff, coarse-textured tree; branchlets upreaching; leaves closely set, not two-ranked; golden yellow in spring and early summer, changing to greenish-yellow and finally to bronze or old gold in late autumn. Found in 1918 on an exposed slope near Eaton, New Hampshire, by Samuel A. Everitt. He published a picture with a brief description on *Garden Magazine and Home Builder,* June 1925. The coloration is best in full sun. This cultivar is sometimes labelled *T. canadensis aurea* in collections.

- - 'Everitt's Aurea Compacta' Hort. See 'Everitt Golden'.

- - **'Fantana'** Swartley ex den Ouden and Boom, Man. Conif. 454. 1965. SPREADING GROUP. Similar to 'Bennett', but the foliage and leaf colour are not quite as good and the plant is usually only slightly broader than high. It was selected about 1913 from seedlings imported from France and propagated by P. F. Avogadro, nurseryman of Bellmore, Long Island, New York. According to Helene Bergman, (*Brooklyn Bot. Gar. Rec.* 21(1): 42. 1965) a plant twenty years old was about 2 feet high and 3 feet wide. The tiered arrangement of the branches and the branchlets in wide, flat sprays, results in a very attractive plant. The leaves (10 mm long by 3 mm wide, reducing considerably in size towards the growth tips) are noticeably broad, elliptic, uniform in shape throughout the plant and always ending in a broad, sharp point; dark green above, stomatic bands beneath, dull in color. Buds hidden by a tuft of very small leaves. The interior of the bush is open and since the main branches are few but their ramifications profuse, the bush appears to be built up of a group of wide, curved ostrich-plume-like sprays arranged gracefully relative to one another so as to produce a most pleasing overall effect suggesting the studied irregularity of a carefully arranged bowl of ferns or similar decorative foliage.

- - 'Fantana Broom' Provisional name (Bergman's Fantana Broom', Raraflora Nurs. Cat. c. 1970. Name only.) Very similar to 'Beaujean'. Propagated by Fred Bergman from a witches' broom on a plant of 'Fantana'. In 1972 the mother plant was 30 inches high and 54 inches wide. If this clone is still in cultivation it should be possible to find a better name.

- - **'Far Country'** New cultivar. GLOBOSE GROUP. This has long been in collections as J.210. I suggest the name to commemorate the late Charles Jenkins who supplied the mother plant to Swarthmore College, Swarthmore, Pa. Layne Ziegenfuss, who obtained material from that college tells me that it is a multi-stemmed plant like 'Innisfree' and 'Youngcone'.

- - f. **fastigiata** See p. 73.

- - **'Fastigiata'** Beissn. Handbuch Nadel. 402. 1891. Beissner described this as an interesting seedling with habit of growth narrow; branches rather short and erect; leaves fern-like in disposition. It was therefore a clone. Identification of any plant now growing in this country with Beissner's plant is impossible, so this name should not be used in any clonal sense. It is safer to regard the clone as lost to cultivation and register any new selection considered to justify introduction under a suitable new name.

- - **'Feasterville'** New name. ('Juniperoides', Raraflora Nurs. Cat. c. 1970. Name only.) A pyramidal hemlock with moderate rate of growth and foliage resembling a juniper. Some leaves on the longer twigs tend to fall off prematurely. The leaves on the inner branchlets are dark green and point forward at about an angle of 45°. It does best in full sun. Originated as a seedling at Raraflora, Feasterville, Pennsylvania, it was formerly listed as 'Juniperoides', an unacceptable name under the *Code.*

- - 'Fernleaf' A listed name. Not worth retention.

- - **'Fremdii'** Jenkins, Heml Arb. Bull. 1. 1932. (as 'Friendi' corrected to Fremdii in Bull. 2. 1933) CONICAL GROUP. An account of the origin of this widely distributed cultivar is given in the Hemlock Arboretum Bulletin 2, dated July 1933. Mr. Frederick D. Fremd, a nurseryman at Rye, New York wrote to Mr. Jenkins: "The parent plant was found by my father (Charles Fremd) in 1887 in the town of Rye. It was first planted in the Rye

T. canadensis 'Detmer's Weeper', in the nursery of Joseph Cesarini, Sayville, Long Island, New York. J.C.S. No. 1828.

T. canadensis 'Curtis Weeping'. Curtis Nurseries, Callicoon, New York, in 1945. J.C.S. No. 1318.

T. canadensis 'Dwarf White-tip', original plant. Morris Arboretum, Philadelphia, Pennsylvania, in 1939. J.C.S. No. 1238.

T. canadensis 'Elm City'. Far Country, Philadelphia, Pennsylvania, in 1945.

T. canadensis 'Ehrle Large-Leaf'. Arnold Arboretum, Jamaica Plain, Massachusetts, in 1970. Accession No. 788-58.

T. canadensis 'Everitt Golden'. Far Country, Philadelphia, Pennsylvania, in 1938. J.C.S. No. 1224.

T. canadensis 'Frosty'. Raraflora, Feasterville, Pennsylvania. J.C.S. No. 1716.

Nurseries, but ten years later, Mr. P. M. Koster of Boskoop, Holland, induced my father to send it to the Koster Nurseries in Holland, whence we later imported many thousands. It forms a compact, pyramidal tree with an irregular ascending main branch system which ramifies into more or less flat twig-sprays in which the terminal shoots everywhere arch over to 0 to +20°. The foliage is very dense right into the interior of the tree; leaves are not completely two-ranked and are dark green". In a later letter, Mr. Koster told Mr. Jenkins that in the first place his interest was because the original tree was the sole survivor of a large block of hemlocks after a severe winter. So it should be a reliably hardy clone.

It was one of the earliest clones to be planted in Britain. Mitchell (Conif. in Brit. Isles. 304. 1972), mentions a tree 34 feet tall at Westonbirt, an arboretum in Gloucestershire. A plant at Far Country, in 1945 at twenty-five years of age, was 12 feet high and 9 feet wide.

- - **'Frosty'** New name (Bergman's Frosty' Raraflora Nurs. Cat. c. 1970 Name only.) WHITE-TIP GROUP. This cultivar is a bushy hemlock of moderate growth with the foliage silvery-white in the summer and the needles minutely tipped with green; flushed with pink and red during the winter. The mother plant in 1970 was 32 inches wide and 36 inches high. In 1972 it was 48 inches wide and 38 inches high. The late Mr. Bergman picked this plant about 1965 from a bed of collected seedlings grown by the author. It is possibly unique—planted in light shade it produces its best colour, almost pure white; in full sun the needles have a tinge of cream.

- - 'Gable' Provisional name. Not now identifiable with any particular clone.

- - 'Gable Cinnamon' Provisional name. Formerly known as Gable Mk 8. "A plant seen by me in the Arnold Arboretum in September 1970 was a hemi-spherical or bun-shaped plant 70 cm across by only half as high with no leader but in its place a well-distributed, rather upright branching system with of course very little leader dominance and all the shoots, down to the final twiglets held at a very narrow (20°) angle of emergence. Older foliage is dark green, tapered and round-ended leaves, but foliage on the current year's growth is a bright grass green above, slightly less bright below but with very inconspicuous stomatic bands. All the growing shoots are carried very irregularly round the shoot, which is sometimes twisted and towards the terminal few cms the leaves become very tiny, few and far between, sometimes absent entirely exposing the shoot, which, arching over consistently in every part of the bush (to ±5°) and being relatively thick and covered with a rich-brown velvety down are very conspicuous. These, with the unusual, yellowish-green leaves gives an overall tone-effect which is quite distinctive. It differs from a plant of 'Cinnamomea' (nearby) considerably in colour and in the very much more prominent brown of the much longer exposed shoots". H. J. Welch. A name more acceptable for registration should be found.

- - 'Gable Dwarf' Provisional name. Formerly Gable Mk. 2.

- - 'Gable Little-Leaf' Provisional name. Formerly Gable No. 1. Layne Ziegenfuss reports a tree of irregular outline, 8 feet high × 8 feet wide at 30 years. The leaves ⅜ inch long by ⅛ inch wide, held in a double rank.

- - 'Gable Twiggy' Provisional name. Formerly Gable No. 5. Layne Ziegenfuss reports a conical tree of very irregular outline 8 feet high × 10 feet wide at approximately 30 years. The leaves were small, 5/16 inch × ⅛ inch, annual growth 4 inches.

More acceptable names should be registered for all the above provisional listings.

- - **'Gable Weeping'** Swartley ex den Ouden and Boom, Man. Conif. 455. 1965. PENDULA GROUP. A slow-growing tree form which may be distinguished from other weeping cultivars by the more compact habit and the much-divided and slender branch structure and the numerous branch tips, all drooping. These give a graceful look to the plant and are of course its means of gaining height. The leaves are narrow and set noticeably close to the shoot, especially towards the tip; 12 mm maximum length, becoming very much less towards the growing tip where there is a cluster of small leaves rather like a child's paint-brush. These, and the pendulous end of every shoot give the plant a graceful look despite the solidity of the foliage. Selected by Joseph B. Gable, Stewartstown, Pennsylvania. The original plant was 4 feet high in 1946. In 1966 it was 17 feet high. This is a distinctive weeping hemlock, widely distributed by the late Henry Hohman, Kingsville, Maryland, as Gable Mk. 3.

- - **'Geneva'** Swartley ex den Ouden and Boom, Man. Conif. 455. 1965. CONICAL GROUP. This clone is named from a tree on the grounds of the Trinity Church Home, 600 Castle Street, Geneva, New York, formerly the home of Peter Maxwell, founder of the famous Maxwell Nurseries, of Geneva, New York. In 1938 it measured 16 feet by 16 feet, being then about 75 years old. In 1973 it was 38 feet high by about 30 feet wide.

'Geneva' makes a beautiful specimen tree or screen. As a young plant it makes an upright, open shrub with very thick, strong, ascending main branches and short laterals in flat sprays. The leaves are variable in length, 5 to 12 mm, very broad, flat, variable in shape from lanceolate to oval, with rounded tips. The branchlets are thick, mid-brown, and the buds also mid-brown, are large and conspicuous. Grafted plants of 'Geneva' are usually narrow for their height, although a specimen at Longwood Gardens, Kennett Square, Pennsylvania, is an exception.

- - 'Gentsch Dwarf Globe' Hort. See 'Palomino'.
- - 'Gentsch Snowflake' Hort. See 'Snowflake'.
- - **'Gentsch White'** New name. ("Variegata Gentsch", Spingarn Amer. Rock Gar. Soc. Bull. **27:** 88. 1969.) WHITE-TIP GROUP. The change in name is necessary for compliance with the *Cultivated Code*. This forms a globose bush in which the older, green foliage is like typical hemlock. The leaves reach 10 mm $\times$ 1.5 mm on older shoots, but towards the tip of the shoot are very much closer together and smaller, averaging 5 mm $\times$ 0.75 mm and creamy-white in colour. The terminal shoots are medium-brown colour, ending in a cluster of white leaves. The whiteness of the foliage is very intense in autumn and holds into winter, especially on plants that are heavily sheared annually. One plant was observed in the collection of Joel Spingarn and two more in the collection of Joe Reis, both of Long Island, New York. Mr. Spingarn's plant at the age of 12 years was 18 inches high and 36 inches wide. This plant burns on the top during most winters. Mr. Reis' plants do not, probably because they are more protected. All these plants originally came from Otto Gentsch, West Merrick, Long Island, New York.

- - 'Gentsch Variegated' Hort. See 'Gentsch White'.
- - '**f. globosa** See p. 73.
- - **'Globosa'** Beissn. Handbuch Conif. Benennung. 65. 1887. In 1891 Beissner (Handbuch Nadel. 402.) described a dwarf, globose, compact plant with the tips of the branchlets pendulous, adding that such somewhat attractive forms turn up here and there in the seed-beds. A plant labelled 'Globosa' was propagated at the Eastern Nurseries, Holliston, Massachusetts, but in all probability it was not the clone described by Beissner, which must now be regarded as lost to cultivation.

Quite a number of these globose plants were recorded by the author in the Thesis of 1939, so it is best now to follow Krüssmann (Handb. Nadel. 343. 1972) and other authors and use the word only as a Group name to encompass all globose clones. See 'Rock Creek' p. 131.

- - **'Globularis Erecta'** (sometimes spelt 'Globosa Erecta') Künkler, Rev. Hort. **56:** 46. 1884. This is another obsolete name for an unidentifiable plant and should no longer be used.

- - **'Golden Splendor'** Mitsch. Nurs. Cat. 11. 1979. GOLDEN GROUP. An upright rather fast-growing golden form which takes full sun better than most clones in this group. It is of unrecorded origin and was introduced by Mitsch Nursery, Aurora, Oregon, 97002. The growth habit is normal except for the colour.

- - **'Gracilis'** Waterer ex Gordon, Pinetum Suppl. 9. 1862. (*Abies canadensis gracilis; Tsuga canadensis gracilis* Carr. Traité Gén. Conif. ed. 2, 249; 'Gracilis Oldenburg' Krüssmann, Handbuch Conif. 343. 1972; NON Hohman, Nurs. Cat. nec Hort. Amer. [in part].)

T. canadensis 'Gable Weeping', outside Henry Hohman's residence, Kingsville, Maryland.

T. canadensis 'Geneva'. Raraflora, Feasterville, Pennsylvania, in 1970. J.C.S. No 1202.

The original plant of 'Geneva', at 600 Castle Street, Geneva, New York, in 1973. Note the difference in dimensions as compared to the tree in the previous photo.

T. canadensis 'Gentsch White'.

T. canadensis 'Gracilis'. Arboretum Trompenburg, Rotterdam. Not to be compared with 'Henry Hohman'.

T. canadensis 'Great Lakes'. U.S. National Arboretum, Washington, D.C. Accession No. 20949.

T. canadensis 'Greenwood Lake'. Far Country, Philadelphia, Pennsylvania, in 1945.

T. canadensis 'Greenspray'. J.C.S. No. 1778. One of the two original plants in Kingsville Nurseries, Kingsville, Maryland, in 1971.

SPREADING GROUP. This name is the subject of much confusion. The original plant as a young plant was described by Gordon as "a very singular looking variety of the Hemlock Spruce, on account of its slender shoots, thin appearance, and small foliage. The leaves are linear, blunt-pointed, glossy above and glaucous below; more or less obliquely placed all round the shoots, and seldom more than three lines (sic.) long. Branches and branchlets very slender, little divided, more or less drooping at the ends, and rather thickly covered with the small, obliquely-placed leaves. A very distinct and singular-looking variety, raised in the nursery of Messrs. Waterer and Godfrey at Knaphill, in Surrey, England".

It soon found a place in French and German Nurseries. Den Ouden and Boom (*Man. Conif.* 456. 1965) considered it as lost to cultivation, but in Welch (*Dwarf Conif.* 321. 1966) there is a picture of a plant which he tells me was taken in an arboretum in Holland. Krüssmann (Handbuch Nadel. 343. 1972) proposed a new name, 'Gracilis Oldenburg' for a clone being distributed by certain German nurseries. He refers however, to a plant 75 years old, from which probably his description was taken. Allowing for the difference in plants at such different ages, I cannot see the distinction he seeks to make, and as Mr. Welch tells me that 'Gracilis' is no longer widely grown in English nurseries there would, on present information, seem to be no good call for departure from the accepted identification of the clone in European production with Waterer's plant. 'Gracilis Oldenburg' is therefore treated here as a superfluous name. Lapses into 'Nana Gracilis' and 'Gracilis Nana' in trade catalogues are nothing more than the mistakes that so often occur, bringing uncertainty, even if not confusion, in their wake.

The use of the name 'Gracilis' by the late Henry Hohman of Kingsville Nursery, Maryland, for a very compact form somewhat similar to 'Hussii' has resulted in confusion of quite another kind. This form was obtained from a German nursery many years ago by the late George L. Ehrle of Clifton, New Jersey, and passed on by him to Mr. Hohman. Despite the trouble his mis-use of the name has caused, it reflects his eye for a good plant, so it is here re-named (see 'Henry Hohman') in his honour.

- - 'Gracilis Nana' See 'Gracilis'.
- - 'Gracilis Oldenburg' See 'Gracilis'.
- - 'Gracillima' A listed name not worth retention.
- - **'Great Lakes'** New Cultivar. This is named for a well-clothed variant in the U.S. National Arboretum, Accession No. 20949. It is a multiple-stemmed tree 10 feet high × 6 feet wide, slow-growing but of loose, upright conical habit with ascending branches and short branchlets. Leaves are dark green, persisting to four years, radially disposed. The source was Ralph M. Warner, Milford, Connecticut.
- - **'Green Cascade'** Vermuelen, Nurs. Cat. 1972. PENDULA GROUP. A strong growing but very dense and compact weeper with pendulous branches. The leaves are more closely set and somewhat shorter than typical hemlock. Selected from a nursery row in the Baier Lustgarten Nursery, Middle Village, Long Island, New York. Introduced by Vermuelen Nursery, Neshanic Station, New Jersey, the original plant was discovered as a 15 inch × 15 inch plant in a block of 4½-5 foot high, normal hemlocks.
- - 'Green Mountain' A listed name. No longer in cultivation.
- - **'Greenspray'** Registered name. Wyman, Arnoldia **23**: 92. 1963. SPREADING GROUP. A compact spreading variant with foliage very similar to 'Bennett', but the leaves do not point forward as much and the tops of the branches are less bare. This clone was selected and registered by the late Henry Hohman of Kingsville Nurseries, Kingsville, Maryland. "Spray-like growths that overlap each growth beneath, the center is open and shows plainly the development of each growth made, which is unlike the mounded forms of dwarf hemlocks. The effect is the development of green sprays, which suggest the name".
- - **'Greenwood Lake'** Swartley ex den Ouden and Boom, Man. Conif. 456. 1965. TWIGGY GROUP. Similar to 'Hussii' but somewhat less dense with less crowded buds; the leaves are very variable in length, from 6 to 12 mm and very broad (2.0 to 2.5 mm), spear-like,

tapering to a sharp point, in two ranks except at the tips of the shoot where they are very tiny, but despite their size, completely hiding the buds. The color is and each leaf has two clear white bands of stomata beneath. Five identical plants were found by a collector near Greenwood Lake, New Jersey, and sold to George L. Ehrle of Clifton, New Jersey. In 1938 the mother plant was 4 feet 6 inches high and 4 feet wide. In 1945 it was about 7 feet high. There is some tendency for this cultivar to sunscald on the top if planted in full sun.

- - **'Guldemond's Dwarf'** Vermeulen, Nurs. Cat. 1964. Conical Group. Originated from D. Guldemond, Longlestown, Pennsylvania. It forms a dense, irregular pyramid. Introduced by Vermeulen Nursery, Neshanic Station, New Jersey.

- - 'Guilford' A listed name. No longer in cultivation.

- - **'Hahn'** New name. Hahn, Gartenswelt. 35: 495. 1931. *(T. canadensis nana)* Spreading Group. This German writer described another variant, using the illegitimate name 'Nana', and included an illustration of a specimen in the rock garden of the Arboretum of Petschau, Bohemia, Czechoslovakia. He briefly stated that the young growth is silver gray, becoming light green, and is about 2 inches long.

- - **'Hancock'** New cultivar. Twiggy Group. A selection from among a number of plants raised from seed of a witches' broom collected near Hancock, New York. The mother plant is owned by James E. Cross, Cutchogue, Long Island, New York, and is about the size of a baseball. It is being propagated and is commercially available. The remaining unselected seedlings from this witches' broom were grown by Joe Cesarini, now of Phytoecology, Ridgely, Maryland. Twenty-seven plants in gallon cans were observed by me in December, 1973. The smallest was about the size and shape of a baseball (3×3 inches), the largest 3 inches high and 6 inches across. The intermediate-size plants resemble 'Minuta', and two others were more regular, giving a clipped appearance. Mr. Cross's plant alone is to be regarded as the clonotype of 'Hancock'. If any other clone is to be introduced, it will need to be registered under a different name.

- - **'Harmon'** Harmon ex den Ouden and Boom, Man. Conif. 456. 1965. Conical Group. Discovered in a block of hemlocks in the LaBar Rhododendron Nursery, Stroudsburg, Pennsylvania and named for Russell Harmon, manager of that nursery for many years. The late Henry J. Hohman, Kingsville Nurseries, Kingsville, Maryland, collected scions in the 1920's and named it 'Harmon'. One of the older grafts in the former Kingsville Nurseries was 6 feet high and 5 feet wide in 1973. The original tree was transplanted to Far Country in November, 1939 at which time it was 8 feet high and less than 4 feet wide. In 1971 it was 17 feet high and 20 feet wide. It is not at all columnar now and has recently lost the clipped appearance it maintained for many years. Young plants are compact in an irregular fashion and are best grown in sun. The lateral branchlets are nearly as long as the terminal.

- - 'Heckman' Provisional name. Listed in a few collections. I have not seen it.

- - **'Helene Bergman'** New name. (Cole's Sport', Helene Bergman, Plants and Gardens, **21:** 43. 1965.) At first a tiny ball of tightly packed foliage with branches gradually lengthening horizontally, forming a low dense mat. Originated as a bud sport on an old plant of 'Cole's Prostrate' in 1957. At fourteen years it was 3 inches high and 9 inches wide. The foliage resembles a dwarf 'Bennett' with indistinct lines of stomata. It shows some bare wood. This clone has been renamed 'Helene Bergman' since there is more than one such mutation from 'Cole's Prostrate' in circulation.

- - **'Heli'** New name ('Bergman's Heli' Raraflora Nurs. Cat. c. 1970. (Name only.) Wide-Leaf Group. An attractive irregular semi-dwarf plant with leaves broader than typical. In 1973 the original plant at Raraflora measured 3 feet in height and 3 feet in width. The foliage is somewhat sparse and open. The leaves are narrow and parallel-sided (up to 15 mm by 1.5 mm, much less on minor shoots) ending usually in a blunt tip. Color is a dull dark green, stomatic bands in conspicuous shoots thin, dark grey-brown, noticeably pubescent the first year, devoid of leaves near the growth tip, so the rather swollen, misshapen buds are prominently seen.

- - **'Henry Hohman'** New name ('Gracilis' SENSU Hohman, Nurs. Cat. c. 1970. NON Waterer). TWIGGY GROUP. The origin of this attractive clone has been given above (See 'Gracilis'). The following is a description taken by Mr. Welch of a plant at Kingsville Nursery in September, 1970.

  - "An upright conical plant 1.8 m high × 0.70 m across with a stout trunk and short (to 30 cm) stout main branches standing up at +30° bearing very short, stiff laterals, so that the plant has a characteristic (and very attractive) rugged outline. Leaves are densely held, 10 × 1.5 mm max. (usually less) very dark green above with the usual white stomatic lines below, but so held, particularly on the projecting shoots, that only the upper side is seen, giving a general effect of dark green. Branches and branchlets are thick, a medium light brown."

- - 'Herman Hesse' A superfluous name—See 'Henry Hohman'.

- - 'Heykoop' A listed name. Not worth retention.

- - **'Hicks'** Swartley ex den Ouden and Boom, Man. Conif. 456. 1965; *atrovirens* Hort. Amer. non Parsons, Nurs. Cat. 1889 nec Beissn. LARGE-LEAF GROUP. In 1973 the writer observed in Highland Park, Rochester, New York, a beautiful LARGE-LEAF hemlock with leaves that were distinctly yellowish green. This plant, labelled 'Atrovirens', was obtained from Hicks Nurseries of Westbury, New York, in 1947 and is now about 30 feet high. It is a pyramidal, slow-growing tree with branches spreading, (differing from 'Macrophylla', in which the branches are erect); leaves densely set and tinged with yellow at all times of the year. Hicks claims to have imported it from France. It is commonly found in collections labelled *T. canadensis atrovirens,* but is renamed 'Hicks' since it cannot be identified as the original 'Atrovirens' of Beissner.

- - **'Hiti'** Wyman, Amer. Nurs. **112** (11): 49. 1960; 'Pomfret' Swartley ex den Ouden and Boom, Man. Conif. 462. 1965; *compacta* Adams (non Sénéclauze) CONICAL GROUP. Pyramidal growing and very hardy with upreaching branches showing the leaves conspicuously bluish beneath. It was found in 1940 in the town of Pomfret, Connecticut, by Ellery B. Baker, then manager of the Hiti nurseries. Sometime later the Adams Nursery, Westfield, Massachusetts, took over the remaining stock and distributed it under the incorrect name 'Compacta'. This cultivar becomes less compact with age.

- - 'Hohman Spreading' See 'Kingsville Spreader'.

- - **'Hornbeck'** Don Smith, Watnong Nurs. Cat. 11. 1972; 'Dr. Hornbeck' Spingarn, Nurs. Cat. 1970 Name only. CONICAL GROUP. A compact, irregular plant with foliage resembling 'Hussii' but slower growing. The original plant is at Arnold Arboretum. It was discovered in Petersham, Massachusetts, by Dr. Peter Hornbeck of the Department of Landscape Architecture, Harvard University, on November 10, 1964, in the woods one-half mile from Harvard Forest. In 1978 the original plant measured 14 inches after fourteen years of cultivation.

- - **'Horsford'** Helene Bergman, Plants and Gardens **21** (1): 41. 1965; 'Horsford's Dwarf' (pro. syn.) Welch, Dwarf Conif. 320. 1966. TWIGGY GROUP. A dense, squat little plant of congested, irregular habit, with ascending main branches and recurving branchlets, annual growth not over 2 inches. The leaves are 10 mm long and proportionately rather broad, spreading at widely differing angles to the shoot, and with a broad band of stomata beneath. Layne Ziegenfuss tells me that on older plants the congestion disappears from the lower part of the plant, where it becomes open with leaves becoming twice the size.

- - 'Horsford Compact' Provisional name. More compact than 'Horsford'. Propagated from a bud sport on 'Horsford' by the late Fred Bergman in 1960. In 1973 it was 8 inches high and 10 inches wide. A more acceptable name should be registered, if the mutation has proved stable and the distinctiveness is worthwhile.

- - **'Horsford Contorted'** Hillside Nurs. Cat. c. 1970 ('Horsford Contorta'); Raraflora Nurs. Cat. c. 1970 ('Horsford Contorted'); 'Pigtail' Hort. Amer. (in part.) There is no doubt that 'Horsford Contorted' is the correct name for this interesting cultivar. It originated with a tree 50-60 feet high (since cut down) found by Wm. C. Horsford in a pasture in Vermont which had slightly twisted branches, contorted branchlets and 'Cinnamomea'-type tips and shown by him to Layne Ziegenfuss and Gregory Williams. It was surrounded by seedlings having similar characteristics. Scions were cut from a selected compact specimen 7 feet 6 inches high and grafted at Hillside Nurseries, Lehighton, Pa. A plant was kindly sent to me and this was shown to the late Fred Bergman of Raraflora Nursery. He later obtained a supply of grafting wood from Vermont and in due time listed it as 'Horsford Contorted' (the change in spelling being made in deference to the present embargo on Latin-form names). But at some point the soubriquet 'Pigtail' was used, and being very descriptive it seems to have stuck. It may be a case where popular usage will prevail, rules or no rules.

It makes an interesting plant, more curious than beautiful. Each year the branchlets twist into tight coils or even knots. Later, as they mature they partially untwist. The tips of the twigs are very pubescent. The leaves are narrow for their length and tend to be acute, in this respect much resembling 'Cinnamomea'.

- - 'Horsford Young Cone' Hort. See 'Young Cone'.

- - **'Horton'** New cultivar. See Stout, Journ. New York Bot. Gar. **40**: 153. 1939. PENDULA GROUP. In an article *Weeping or Pendulous Hemlocks'* the late Dr. A. B. Stout first drew attention to what was the largest and so presumably, the oldest known specimen of weeping hemlock. Since the exact spot where General Joseph Howland is reputed to have found the four original 'Sargentii' seedlings is not on record, the relationship of the Horton tree to the others cannot be established, but a direct connection is at least probable from geographical and other considerations, and the characteristics of the tree described by Dr. Stout are close enough to justify our acceptance of the Horton hemlock as the clonotype one of the (now five) clones which comprised the cultivar 'Sargentii' as originally circumscribed.

- - **'Howe'** New name. Named from a compact, conical plant growing on the Howe, Pennington, N.J. that was found in Everton, New Hampshire and introduced by Hess Hawken's, Wayne, N.J. under the illegitimate name 'Howei'.

- - 'Hubble'. A listed name. No longer in cultivation.

- - 'Huffmann' Provisional name. Listed Raraflora Nurs. Cat. c. 1970. In a few collections but I have not seen it.

- - **'Hunnewell'** New cultivar. PENDULA GROUP. The clonotype is the large specimen in the Wellesley Arboretum, Massachusetts, New York. The case for doubting that this is the original Sargent hemlock collection No. 3 (see p. 135) seems to rest on the omission of any reference to this tree (despite its appearance on a map drawn at the time and the tree-like proportions it would undisputedly have attained by 1895) in the writings of H. H. Hunnewell or his cousin, C. S. Sargent, and the statement by M. Hornibrook in 1923 that the tree was dead. Of all errors to which writings of this kind are prone, omissions are by far the most common: Hornibrook's statement was unsupported, and its repetition *ad nauseum* by later writers adds nothing to its authority. Del Tredici (*A giant among the dwarfs* 42. 1983) is right in saying that certainty on the question is not possible, but in my judgment such meagre evidence as is available, and the tree itself point to such convincing probability of this being the original tree that its selection in this way is justified.

- - **'Hussii'** Jenkins, Heml. Arb. Bull. 1. 1932; 'Hussi' Gotelli, Wm. T., Amer. Hort. Mag. **39** (1): 186. 1960. TWIGGY GROUP. A very slow-growing upright shrub, eventually a small tree of dense, irregular habit rarely with definite terminal shoots, characterized by very crowded, short and twiggy branchlets, with some branch tips fasciated. Annual growth 1-2 inches. Leaves semi-radially disposed, densely crowded to 8 mm × 1.75 mm on older branches, much less on new shoots. Shoots yellowish-brown, thick, minutely pubescent. Buds swollen and prominent, not hidden by a cluster of tiny leaves around the terminal bud. 'Hussii' usually develops an open habit with age. Found in 1900 by John F. Huss, then Superintendent of Parks at Hartford, Connecticut. The plant at Far Country was 1 foot 7 inches high in 1940. In 1945 it was about 3 feet 4 inches high. In 1970 it measured 17 feet.

T. canadensis 'Harmon'. Far Country, Philadelphia, Pennsylvania, in 1945. The original plant. J.C.S. No. 1440.

T. canadensis 'Henry Hohman'. U.S. National Arboretum, Washington, D.C. Accession No. 25750. Plants in nurseries that have been cut into for propagation material are usually denser in apparent growth.

T. canadensis 'Helene Bergman'.

T. canadensis 'Henry Hohman', outside Henry's own front door at Kingsville, Pennsylvania.

T. canadensis 'Hornbeck'. Arnold Arboretum, Jamaica Plain, Massachusetts, in 1970. J.C.S. No. 1824.

T. canadensis 'Hicks' (syn. 'Atrovirens'). Far Country, Philadelphia, Pennsylvania, in 1945.

T. canadensis 'Horsford'. U.S. National Arboretum, Washington, D.C.

T. canadensis 'Hussii'.

- - 'Ideal' Hort. See 'Curtis Ideal'.
- - 'Imperial New cultivar. WHITE-TIP GROUP. More compact than typical hemlock, with the white tips fading to a cream color. Distributed by Brimfield Gardens Nursery, Wethersfield, Connecticut.
- - 'Innisfree' New cultivar. Listed Hillside Nurs. Cat. c. 1970. (Name only.) GLOBOSE GROUP. A multi-stem globe discovered by Alfred Fordham, propagator at Arnold Arboretum, in the town of Innisfree, Massachusetts, in 1950 (Accession Number 1026-50). Growth similar to 'Young Cone', with main branches stiffly upright, nodding at the tips.
- - 'J.210' See 'Far Country'.
- - 'Jacqueline Verkade' Registered name, Verkade ex Wyman, Arnoldia **29:** 8. 1969. CONICAL GROUP. The following information was supplied by John Verkade of Verkade's Nurseries, Wayne, New Jersey:

> This plant was found in our seedling beds in 1961, when three years old. It has an average growth of one-half to three quarters of an inch per year. It has a perfect conical form and is very dense, with dark green foliage throughout the year. This plant is now (1968) 5½ inches tall and is estimated to be ten years old. It will grow in full sunlight and will not burn in summer or winter.

The leaves are quite small, 3-6 mm long and quite scattered, especially at the tips of the twigs. This cultivar is unique in that there are scarcely any leaves that are reversed and appressed to the top of the branchlets. The foliage is suggestive of *Thuja orientalis* 'Rosedalis Compacta', although it does not undergo the change in color characteristic of that little plant.

- - 'Jan Verkade' New cultivar. SPREADING GROUP. This arose as a seedling in the Verkade Nursery at Wayne, New Jersey. As in 'Ashfield Weeper' and 'Kelsey's Weeping', a single shoot grows off to one side, but unlike those two cultivars this creeps along the ground unless staked. The leaves of the summer growth are nearly radial in their distribution; blunt, short, 5-10 mm long. In contrast the needles on two-year old wood are acute and very long, up to 19 mm. The foliage is softer to the touch than that of Kelsey's Weeping'. It is a fast grower and has been propagated.
- - 'Jeddeloh' Grootendorst, Dendroflora **2:** 49. 1965. SPREADING GROUP. A dwarf spreading hemlock with an open center resembling a birds nest; slow growth with fresh green needles make it an outstanding subject for suburban landscaping. This clone has been widely distributed in Europe by the Jeddeloh Nursery, Oldenburg, West Germany, so this name will doubtless survive as a commercial synonym.
- - 'Jenkinsii' Bailey, Cult. Conif. 124. 1933. LITTLE-LEAF GROUP. A fast-growing pyramidal tree with an open, ascending branch habit. The original plant was a seedling in the Towson Nurseries, Towson, Maryland. When about 2 m high, the late C. E. Jenkins wrote (Heml. Arb. Bull. 1. 1932) that it was "the most graceful and unusual type in the collection." It does not retain this quality in maturity, but as a young plant it is outstanding. The leaves are shorter and finer even than 'Microphylla'.
- - 'Jenkins Nana' A listed name. Not traced in cultivation.
- - 'Jennings Yewlike' New cultivar. YEW-LIKE GROUP. An upright, fast-growing form with leaves quite two-ranked, somewhat greater in thickness and length than typical hemlock and markedly acute near the tips of the branchlets. The mother plant, now at Far Country, was a seedling selected at the F. & F. Nursery, Princeton, New Jersey, by John J. Jennings. The plant was less than 3 feet high in 1935. In 1970 it was 15 feet high and about the same breadth. It is intermediate in rate of growth between 'Taxifolia' and 'Valentine', keeping a good compact habit without trimming.
- - 'Jervis' Helene Bergman, Plants and Gardens **21**(1): 43. 1965; 'Nearing' Swartley ex den Ouden and Boom, Man. Conif. 460. 1965. TWIGGY GROUP. This was discovered by the late G. G. Nearing of Metuchen, New Jersey. A layered branch was grown on by

Edward Thuem also of Metuchen and in 1956 the late Fred Bergman rooted cuttings of this plant and distributed it as 'Jervis' (after Port Jervis, New York). Both names have the same publication date, but since this cultivar is now widely known as 'Jervis' this name is preferable. It is similar to 'Hussii', but it maintains a more compact habit as it grows older.

This is now a very popular variety and quite large plants of 'Jervis' (pronounced 'Juvviss' not 'Jahrviss') can be seen. These are very densely clothed with congested dark-green foliage. The shape seen from some distance away is regularly conical, but closer inspection shows the outline of the bush to be broken up by the numerous main branches projecting beyond the general line by as much as 7 cm and everywhere standing up stiffly at 80°. These shoots bear the usual crowded foliage at their tips and usually also along their length, as do the less prominent shoots that do not thus project, and all this foliage gives a dense and impenetrable look to the plant which is characteristic. The leaves are long-oval, pointed at the tip. Although uniform in shape they vary widely in size, from 3 mm long in the tufts (where they are very close and congested) to 10 mm by 2 mm on the main shoots; dark green both above and below, where two narrow bands of stomata close to the midrib are quite inconspicuous, so this clone appears to be one of the darkest green of any. The buds are visible but inconspicuous.

Some years ago the late Fred Bergman noticed a bud-sport on a branch of 'Jervis'. This he propagated, calling the resulting plants 'Jervis Sport', claiming that it made twice the annual growth of the mother-plant. Spingarn (Amer. Rock Gar. Soc. Bull. **27** (3): 87. 1969) stated that he could see no difference. Doubtless, time will tell which is right.

- - **'John Swartley'** New name. ('Dr. Swartley'. Hort. Amer.) SPREADING GROUP. A slow-growing spreader with branchlets fan-shaped and somewhat crowded; leaves a little shorter than typical, medium yellow cast in winter, and with dull lines of stomata beneath. It was discovered several years ago by Greg Williams near Weissport, Pennsylvania. The original plant is in the collection at Palette Gardens, Quakertown, Pennsylvania. In 1975 it was about 10 inches high and 18 inches wide. The name change is in compliance with the *Cultivated Code* recommendation against using forms of address in cultivar names.

- - 'Julians' See 'Slenderella'.

- - 'Juniperoides' Hort. and 'Juniperlike' Hort. See 'Feasterville'.

- - **'Kathryn Verkade'** Verkade ex Spingarn, Amer. Rock Gar. Soc. Bull. **27** (3): 89. 1969. (As Verkade Witches' Broom.) SPREADING GROUP. Spingarn describes it as follows:

> A very distinct form derived from a witches' broom by Verkade's Nurseries in Wayne, New Jersey. It somewhat resembles the form *T. canadensis* 'Bennett' but in a miniaturized version. The plant is horizontally tiered, flat-topped and very symmetrical. It requires a somewhat shaded site. The average growth of an older plant is 2½ inches. Younger plants start slowly. An eight-year-old plant is 6 inches high and 15 inches across.

Mr. Verkade is of the opinion that Illustration No. 460 in *Manual of Dwarf Conifers* Welch, 477. 1979 is of this cultivar, (not 'Jacqueline Verkade' of the caption in the book.) He tells me that in the early days there was inadvertent confusion in the labelling on the nursery, so there may be plants in collections wrongly labelled. The descriptions are clear. The name is changed at Mr. Verkade's request.

'Beaujean', 'Fantana Broom', 'Greenspray', and 'Narragansett', are all very similar in habit, foliage, and rate of growth, but they are all separate clones.

- - **'Kelsey's Weeping'** Swartley ex den Ouden and Boom, Man. Conif. 458. 1965; 'Kelsey's Weeping Hemlock' Kelsey, Horticulture. **26**: 42. 1948. PENDULA GROUP. An asymmetrical form (curious but not without a bizarre beauty in a spot where it has room to develop its peculiar habit of growth), combining a single, sloping, long-reaching stem with long, pendulous branches. As the curious stem elongates it usually lifts up higher every year, then after a while the tip branches curve downward. According to Stout (Journ. New York Bot. Gar. **40**: 162. 1939) four similar plants were discovered in 1929 on the property of H. P. Kelsey, nurseryman of East Boxford, Massachusetts. According to Mr. Kelsey, one was given to the Arnold Arboretum, two were planted in the nursery stock block, and one was given to Hunnewell Arboretum, Wellesley, Massachusetts. One on the nursery grounds and the one in Hunnewell Arboretum soon died. The plant at Arnold Arboretum (at one time erroneously labelled 'Pendula') is still there.

- - 'Kingston Hollow' New cultivar. LITTLE-LEAF GROUP. A bushy little tree with very unusual texture; most leaves are short, 3-5 mm, with a few up to 10 mm in length; the majority of the leaves are twisted and aborted, especially on the spring growth. The tops of the branchlets are very bare looking and unusually pubescent. Walter Kolaga, proprietor of the then Mayfair Nursery, Nichols, New York, found this plant on his own property in about 1970. In 1973 it was 3 feet high, and it is now in the collection at Hillside Nursery, Lehighton, Pa.

- - 'Kingsville' Hohman, Nurs. Cat. 10. 1946; Hohman ex Wyman. Amer. Nurs. **112**(11): 12. 1960. FASTIGIATE GROUP. The original plant in the Kingsville Nurseries, Kingsville, Maryland, was very narrow, measuring 14 feet high and 4 feet wide in 1945. It had been cut back hard for scion material year after year. The writer has observed grafted plants in several collections. Without clipping the plants grow as wide as typical hemlock but the branches are widely spaced. It would appear to lack sufficient distinctiveness to justify retention.

- - 'Kluis Pyramid' A listed name. I have not traced it in cultivation.

- - 'LaBar Gem' Swartley, Amer. Nur. **133**(7): 7. 1946. CONICAL GROUP. Pyramidal, compact, rather stiff, about as broad as high, rather slow-growing; leaves densely set, not two-ranked, dark green. This cultivar has enough irregularity to make it interesting. It was collected in West Virginia in 1936 by the LaBar Rhododendron Nursery, and in 1938 was transplanted to their showgrounds in Stroudsburg, Pennsylvania. In 1938 the original plant was 4 feet high and a little less in width. In 1948 it was 7 feet 4 inches high and 6 feet 6 inches wide.

- - 'LaBar White-Tip' Swartley, Amer. Nurs. **133**(11): 10. 1946. WHITE-TIP GROUP. This is a free-growing, irregular, bushy tree displaying more whiteness than many other clones in this group. It was observed in the LaBar Nursery in 1945, measuring 4 feet in height and a little more in breadth. It was transplanted to Far Country, and in 1970, was 23 feet high and 24 feet wide. It is a clone, so should not (where its identity is known) be distributed under the loose name 'Albo-Spica'.

- - f. **latifolia** See p. 73.

- - 'Latifolia' Sénéclauze, Conifères. 19. 1868. Sénéclauze described this clone as a beautiful variety, as vigorous as the type, distinct and very remarkable; found in their nursery fields at Bourg-Argental, Loire, France in 1867. He described the leaves as not exactly two-ranked, 8-10 mm long and 2-3 mm wide, scarcely narrowed at the base. Less graceful habit than normal, with shorter much broader leaves. Propagated by John Mitsch, Aurora, Oregon.

- - 'Laurie' Swartley, Amer. Nurs. **133**(7): 10. 1946. GLOBOSE GROUP. Spherical or oval in outline with no leader, the structure consisting of numerous erect branches, green with slender branches and branchlets. This cultivar, when grafted, has a less compact habit and inferior foliage. It will probably grow to 6 feet in fifteen to twenty years. It was selected by Robert Laurie, Proprietor of the Stoughton Nurseries, Stoughton, Massachusetts, from a bed of collected seedlings. About 1935 three hundred plants were rooted from cuttings. Nine years later they were 2-3 feet high.

- - 'Levi.' Provisional name. A variant found in Milton, Pennsylvania, by one of the late Fred Bergman's neighbours. It forms a compact, pyramidal plant with dark green leaves, 4-20 mm long, scattered, some twisted. I do not know whether it is still in cultivation.

- - **'Lewis'** Helene Bergman, Plants and Gardens, **21**(1): 43. 1965. TWIGGY GROUP. This cultivar, selected by Mr. Clarence McKinley Lewis of Skylands, Sterlington, New York, was sent to Mr. Carl Starker of Aurora, Oregon, and was propagated by the Mitsch Nursery in Aurora, Oregon, who at first sent it out under the name 'Dwarf Upright'. It is a beautiful, slow-growing, upright variant, forming a very irregular pyramid with rather stiff and rigid growth; branches thick, light-brown, pubescent, somewhat sparse but branchlets and leaves crowded. Leading shoots prominent, with erect, tightly appressed leaves. At twelve years a plant at Raraflora, Feasterville, Pennsylvania, was 27 inches high and 20 inches wide.
- - 'Lincoln White-Tip' A listed name. I have not found it in cultivation.
- - **'Little Joe'** New cultivar. TWIGGY GROUP. Obviously a seedling, but the origin is not on record. It forms a tight, congested bun. The leaves are smaller than in 'Minuta' and the annual growth possibly a little less and the plant forms a smoother mound. It is one of the smallest clones known. At 20 years a plant on the Mitsch Nursery, Aurora, Oregon is 6 inches high and 10 inches across.
- - 'Longwood' See 'Slenderella'.
- - 'Loudon' A listed name. I have not found it in cultivation.
- - **'Lustgarten Creeping'** Vermuelen, Nurs. Cat. 1971. PENDULA GROUP. This clone originated in the Baier Lustgarten Nursery, Middle Village, Long Island, New York. It is more pendulous and slower in growth than the Sargent hemlock. It has been propagated by John Vermuelen and Son, Neshanic Station, New Jersey, and Joe Cesarini, now of Ridgely, Maryland. The latter had a plant 28 inches high and 30 inches wide at thirty years.
- - 'Lutes' Provisional name. Listed Brimfield Gardens Nurs. Cat. 30. 1967. This was described as being of "brilliant gold" colour, but neither this nor 'Lutescens', another Provisional name, listed by Hillside Nurs. Cat. c. 1970 (Name only) seem to have survived as relating to any particular clone. Both names are illegitimate so should now be regarded as lost to cultivation.
- - f. **macrophylla** See p. 73.
- - **'Macrophylla'** (Beissn.) Fitschen, Handbuch. Handb Nadel. ed. 3, 72. 1930. *T. mertensiana* f. *macrophylla* Beissn. Handbuch. Conif. 402. 1891. LARGE-LEAF GROUP. This was described as a quick-growing, vigorous form of French origin with long, wide leaves. Plants of this description were observed in 1938 in several places. One, at the T. A. Havemeyer Estate, West Grove, Long Island, New York, about 13 feet high, another in the Bayard Cutting Arboretum near Oakdale, Long Island, New York, which was about 25 feet high, but neither tree could be found in 1970. Another was seen at the nursery of Jacob C. van Heiningen, 60 Danbury Road, Wilton, Connecticut. In 1945 it was about 18 feet high. Mr. van Heiningen said it had come from Parson's Nursery on Long Island about 1920, when it was about 4 feet high. In the 1930's Mr. van Heiningen sent several plants of the same kind to Mr. Jenkins at Far Country. Two similar plants labelled 'Macrophylla' were obtained from Hillier and Sons in 1938 by Mr. Jenkins. The larger one, in 1970, measured 22 feet high and 19 feet wide. Both of them remain in the collection at Far Country.

  All that can now be said is that none of these plants has any claim to being the original French plant, so 'Macrophylla' has become a very loose term and should no longer be used in any clonal sense.
- - **'Mansfield'** Swartley, Amer. Nurs. **134**(11): 10. 1946. GLOBOSE GROUP. In 1945 the author observed some of a dozen or so seedlings that had been discovered by Dr. Raymond Wallace in a wood in Mansfield owned by the University of Connecticut. They were all tall globes with seven or eight stems, each sub-dividing at a sharp angle. No mother-tree could be found, so probably this variant comes true from seed.
- - 'Many Cones' and 'Many Stems' I have been unable to trace in cultivation. They seem to be no more than collectors' nicknames.

T. canadensis 'Jeddeloh'. Jeddeloh Nursery, Oldenburg, Germany. Photo from H.J. Welch.

T. canadensis 'Jacqueline Verkade'. Collection of Joel Spingarn, Baldwin, Long Island, New York, in 1973. J.C.S. No. 1759.

T. canadensis 'Jenkinsii'. Far Country, Philadelphia, Pennsylvania, in 1948. J.C.S. No. 1234.

T. canadensis 'Jervis'.

T. canadensis 'Kelsey's Weeping'. Fowle's Nursery, Newburyport, Massachusetts. Photograph supplied by Arnold Arboretum, Jamaica Plain, Massachusetts.

T. canadensis 'Kingston Hollow'. Discovered on the land of the former Mayfair Nursery, Nichols, New York.

T. canadensis 'Lewis'. Collection of Joel Spingarn, Baldwin, New York, in 1973. J.C.S. No. 1713.

T. canadensis 'LaBar White-tip'.

- - **'Matthews'** Jenkins ex den Ouden and Boom, Man. Conif. 459. 1965; *T. canadensis* var. *Matthewsii* Jenkins, Heml. Arb. Bull. 11. 1935. GLOBOSE GROUP. This clone forms at first a compact bush, eventually losing its semi-spreading or globose habit; leaves a little longer and broader than typical hemlock. Found in 1933 in Pike County, Pennsylvania, by Edwin Matthews of the Outdoor Arts Nurseries, Philadelphia, Pennsylvania. Jenkins described it as a seedling originating in Vermont, but this was an error. The foliage is somewhat like 'Bristol'. The original plant is at Far Country, measuring 17 feet high and 16 feet wide in 1970. Propagation has been limited.
- - 'Mehalic's Dwarf' Hort. See 'Mihalic's Dwarf.'
- - 'Melville' A listed name. I have been unable to trace this in cultivation.
- - 'Merritt's Golden' Probably not in cultivation if it ever was distinguishable from 'Everitt Golden'. It might well be merely a mis-spelling. These faded labels leave a lot to answer for!
- - **'Meyers'** Swartley ex den Ouden and Boom, Man. Conif. 460. 1965. CONICAL GROUP. Pyramidal and compact; branches heavier than usual, with lateral buds on the twigs developing more freely; leaves rather broad, dark green, resistant to windburn. Found in north-eastern Pennsylvania by E. W. Meyers, nurseryman, Hatboro, Pennsylvania.
- - f. **microphylla** See p. 73.
- - **'Microphylla'** *T. canadensis microphylla* (Lindl.) Sénéclauze, Conif. 20. 1868; *Abies canadensis microphylla* Lindl. in Gar. Chron. **1864**: 460. 1864. LITTLE-LEAF GROUP. Lindley was the first to describe var. *microphylla*: "An Abies of such singularly dwarf habit that it might be almost compared to a heath; leaves dark green, with a white streak beneath, rough at the edge, and no bigger than those of *Menziesia polifolia*." This epithet is often used as a cultivar name or loosely in a collective sense (see, for example, Krüssmann, Handbuch Nadel, 344. 1979). There are several clones of divergent origin being propagated and distributed as 'Microphylla' although only the clone described by Lindley is entitled to the name. Other clones worth distinguishing should be registered under new names.
- - 'Mihalic's Dwarf' A listed name. I have not found it in cultivation.
- - **'Milfordensis'** *T. canadensis* var. *milfordensis* (Young ex Gord.) Nich. Dict. Gar. 101. 1887; *Abies canadensis milfordensis* Young ex Gord. Pinetum 421. 1875. LITTLE-LEAF GROUP. According to Gordon, it is "a dwarf variety, globular in form, with the shoots slender and drooping, and the leaves much smaller than common Hemlock Spruce. It is quite distinct from var. *gracilis* and it originated in the nursery of Mr. Maurice Young at Milford in Surrey."

    According to a letter from Sir Arthur Hill, then Director of the Royal Botanic Gardens, Kew, Richmond, Surrey, England, an herbarium specimen there of "Abies canadensis dwarf variety (Young's)" was received in 1878. This is possibly 'Milfordensis'. Rothrock (1880, p. 15) mentioned 'Milfordensis' in his record of plants growing near Horticultural Hall in Philadelphia. This name was also listed in the Kew Handlist of Conifers (1925), but according to a recent letter from Kew it is no longer in the collection there.

    'Milfordensis' must therefore be regarded as lost to cultivation, unless a 10 foot plant at Kew, mentioned by A. F. Mitchell (Conif. Brit. Isles. 304. 1972) can substantiate its claim to the name.
- - 'Miller' A listed name. I have not found it in cultivation.
- - 'Millstream' A listed name. Probably not in cultivation.
- - **'Minima'** Hesse ex Schelle, Winterhart Nadel. 114. 1909. SPREADING GROUP. "Dwarf shrub, very slow-growing; branches spreading, branchlets very short; leaves 5-8 mm long, much shorter on some branchlets at base and tips, green above, with two glaucous green bands beneath, giving the plant a greyish appearance."

    There is no doubt that this German clone (or at least what passes for it in this country) and our American-based 'Bennett' are very close, so close indeed that had their separate origins, continents apart, not been on record they would have almost certainly have been treated as being a single cultivar.

Welch (Dwarf Conif. 321. 1966) quotes information received from Mr. Gotelli of East Orange, N.J. "A truly fine form which is suitable for use as an individual specimen. It is also suitable for a large rock garden as it will ultimately develop into a fair-sized plant. My oldest specimen was 1.25 m high with a spread of 3 metres. One of the real assets of this form is that it does not lose its shape. It was perhaps this that prompted the late Henry Hohman to tell the author that he agreed that 'Minima' had the edge on 'Bennett' and then to add "But one must be careful not to get the labels mixed."

There is however, one other possibility. The picture given by den Ouden and Boom (Man. Conif. 459. 1965) as 'Minima' is certainly not the same plant. I have been unable to find any record of how or where the Gotelli plant reached him, nor come to that, of any importation of 'Minima' into America. So it may be that 'Minima' of American gardens is not the German clone.

- - **'Minuta'** Teuscher, New Flora and Silva **7**: 274. 1935 ("var."). TWIGGY GROUP. Teuscher described the plant: "A dwarf, compact plant of somewhat irregular shape, but about as broad as it is high. Annual increase in growth not more than 1 cm. The oldest plant which is known and which produces cones and fertile seeds is estimated to be more than fifty years old but does not reach more than two feet in height. Its needles—1 to 1.5 mm broad and 6 to 10 mm long—are dark green above and marked with two conspicuous lines beneath. This variety breeds true from seed".

Accounts of the origin of this plant are badly jumbled. It is therefore necessary to lay to rest the confusion that has arisen between 'Minuta' and 'Abbott's Pygmy'. They are somewhat similar in appearance, but not identical either in characteristics or origin. They differ in rate of growth and the size, shape and colour of the leaves.

'Minuta' was brought into circulation by George L. Ehrle, a florist and nurseryman in Clifton, New Jersey. A Mr. Craig informed Mr. Ehrle of the existence of this very dwarf strain and also told him that Daniel M. St. George of Charlotte, Vermont, was the discoverer. Mr. St. George wrote that two plants were available for five dollars each, so Mr. Ehrle quickly ordered them and they were received September 10, 1934.

Mr. St. George, in a letter to Mr. Ehrle, related his activities as follows: In 1927 he dug some dwarf hemlocks on a north slope of the Green Mountains, at a rather high altitude in a small open area with normal hemlock growing on all sides. In 1931, and again in September 1934, he dug more plants. On this later visit he found the "parent", about 2 feet high and bearing cones. In the spring of 1935 he again visited the site and found not a single small plant; even the parent was dead—apparently destroyed by grazing cattle. Altogether he claims he dug and sold twenty-five to thirty plants. Mr. Ehrle chopped up one plant for scion material, and the other was pictured and described as var. *minuta* by Henry Teuscher in *New Flora and Silva*.

Mr. Teuscher, in this, the first published account of 'Minuta', gives Mr. Ehrle credit for calling this dwarf variety to his attention but does not name the finder. His account is essentially accurate. He wrote: ". . . returning to the same place several times during the last eight years, he (the finder) collected altogether some twenty-five seedlings, of which all are dwarf and show surprisingly little variation."

Jenkins (Heml. Arb. Bull. 62; 1948) initiated the confusion when in describing the discovery of 'Abbott's Pygmy' by Frank L. Abbott he added, "In the old days before Standardized Plant Names was issued we called it *T. canadensis minuta*." He emphasized this confusion by inserting a picture of 'Minuta' (not from the present author) very similar to the print originally supplied by George L. Ehrle and used by Mr. Teuscher in 1935. Jenkins at once corrected his mistake (Hem. Arb. Bull. 63: 1948) but unfortunately it would seem that Teuscher did not see this latter publication, for later he wrote (*Plants and Gardens*—New Series **5**(3): 141. 1949) an article in which both plants and their respective histories are inextricably mixed up. Mr. Welch (Man. Dwarf. Conif. 371. 1979) sums it up by telling us: "The heading relates to one plant, the sub-heading relates to the other, and confusion between Daniel St. George and Frank L. Abbott turns it all into nonsense."

In a letter to the author, Mr. Teuscher stated he had sent a questionnaire to Mr. Ehrle, but as he received no reply wrote the article in *Plants and Gardens* based on Mr. Jenkins' earlier account. In an article in the American Nurseryman, Clarence F. Lewis (1961) also mistakenly gives the credit to Frank L. Abbott for discovering 'Minuta'.

To summarize: 'Minuta' was discovered by a plant collector, Daniel M. St. George of Charlotte, Vermont. Over a period of years Mr. St. George dug and sold about twenty-five of these dwarf plants found near a 2 foot high plant bearing cones. A Mr. Craig learned about these dwarfs and informed George L. Ehrle, a florist in Clifton, New Jersey. Mr. Ehrle purchased two plants, told Mr. Teuscher about them and sent him a plant and a photograph in a strawberry box.

'Minuta' is reliably dwarf and is now one of the most popular clones in its group.

- - 'Minuta Pendula' Hort. See 'Starker'.

- - 'Minuta Variegata' An unstable sport no longer in cultivation.

- - **'Moll'** Krüssmann, Nadel. 288. 1955. CONICAL GROUP. Krüssmann described this cultivar as bushy and compact with a loose arrangement of branches, leaves close together. It was discovered prior to 1939 and was named after its finder Peter Moll of Heisterbacherrot, Germany.

- - **'Moon's Columnar'** Hunnewell ex Jenkins, Heml. Arb. Bull. 14. 1936. FASTIGIATE GROUP. The late Charles Jenkins records a tree at the Hunnewell Arboretum at Wellesley, Massachusetts, and states, "It is (1936) about 20 feet tall, very dense and distinctive." It appears to be lost to cultivation.

- - 'Muttontown' Provisional name. The difficulty about this name and another name 'Christie' sometimes met with (which is clearly a synonym of 'Muttontown') and their relation to 'Armistice' has been discussed under that cultivar. Mr. Charles Jenkins is on record as having sent a plant purporting to be 'Armistice' to Mr. Marshall of Brimfield Gardens. This plant at some later date was transferred to the Christie Estate and from there was propagated by two nurserymen, one of whom distributed plants as 'Muttontown' and the other as 'Christie'. So far so good. But if there were (as I think probable) either two clones not distinguished by Ralph Warner, or sufficient variability to amount to the same thing, Jenkins' plant might well have been different to that purchased by Wm. T. Gotelli from Warner in 1957. In such a case, the plant in the collection at Christie Park, Nassau County, Long Island, New York is available as a clonotype for 'Muttontown' The two plants are similar in appearance, and both rather variable from plant to plant, forming irregular, squat to upright compact plants, but 'Muttontown shows more irregularity in growth and foliar characteristics, many of the leaves being longer, up to 10 mm. Some critical examination of the plants in cultivation is required to dispose finally of this problem. One fact that would confirm the existence of two clones would be the absence of any growth with intermediate characteristics.

- - **'Nana'** Carrière, Conif. 190. 1855. SPREADING GROUP. As might be expected in a species that gives birth to plants an outstanding feature of which is abnormally slow growth, this name (*nana* is the Latin word for "small") has been used by many authors for plants of this description that have come to their notice. These have, however, always been clones—the characteristic is undefinable and it is associated so frequently with other deviations from the normal that it does not lend itself to inclusion in any botanical classification.

Carrière described his var. *nana* as "a bushy shrub, often recumbent or spreading." Webster (Hardy Conif. Trees 13. 1896) tells us his plant was "of very dwarf and spreading growth with short branches and closely-set tufted leaves." Hornibrook (Dwarf Conif. 185. 1923) describes a form with unattractive yellow-green foliage. Hoopes (Book of Evergreens 188. 1868), Gordon (Pinetum 15. 1858) and later writers swell the chorus, but what is certain in every case is that the clone being described will no longer be identifiable in cultivation today. This effectively proscribes the use of the epithet nowadays in any clonal sense. Any dwarf clone in cultivation as 'Nana', should be registered under a new name. See 'Hahn'.

- - 'Nana Gracilis' Hort. See 'Gracilis'.

- - 'Nana Pyramidalis' An unacceptable name listed by Raraflora, c. 1970. I do not know whether it is still in cultivation.

- - **'Narragansett'** Registered name. Brandley ex Hebb, Arnoldia **32**(6): 287. 1972. SPREADING GROUP. This was given the temporary name of 'Brandley Broom' at the Arnold Arboretum. It is similar to 'Beaujean'. James W. Brandley, Pawtucket, Rhode Island, discovered the original as a witches' broom growing on a plant of typical *T. canadensis* in Providence, Rhode Island in 1958. According to Mr. Brandley, the original witches' broom was round, low-spreading, and concave in the center. It was approximately 2 feet in diameter and about twenty years old. It was growing on a branch about 10 feet from the ground in dense shade. Two years later it died. A plant at the Arnold Arboretum, Accession No. 80-64, was propagated in 1964. In 1970 it measured 12 inches high and 24 inches wide.

- - 'Nearing' Swartley ex den Ouden and Boom, Man. Conif. 460. 1965. See 'Jervis'.

- - 'Newport' Provisional name. Listed Spingarn Nurs. Cat. c. 1967 (Name only.)

- - **'Outpost'** Swartley, Amer. Nurs. **83**(8): 11. 1946. PENDULA GROUP. A gracefully weeping plant with excurrent habit. Fast-growing plant; branches more or less pendent, branchlets quite pendulous, leaves not quite two-ranked. Den Ouden and Boom (Man. Conif. 461. 1965) state that it was found in a seed-bed of the J. M. Woodcock Company in 1934, but the author saw the original plant in 1938 and it then measured 10 feet high. In 1945 it was about 18 feet high.

- - **'Palomino'** Spingarn, Amer. Rock Gard. Soc. Bull. **27**(3): 88. 1969; 'Gentsch Dwarf Globe' Spingarn, Plant list 1973. CINNAMON-TIP GROUP. A very slow-growing variant from Otto Gentsch, West Merrick, Long Island. The mother plant in 1970 was 13 inches high and 21 inches wide, about twenty-one years old. This plant forms a very dwarf compact bush with irregular, congested growth. It having been in circulation at one time under that name, 'Gentsch Dwarf Globe' may be regarded as a commercial synonym.

- - 'Parkeri' Provisional name. Listed Mayfair Nurs. c. 1973.

- - 'Parsons Little-Leaf' Provisional name. LITTLE-LEAF GROUP. This name is suggested for a tree with short leaves growing at the Arnold Arboretum (Accession No. 1508A) a grafted plant received from Parsons and Sons in 1880. In 1938 it was 28 feet high and 20 feet wide. It grows a little more than half as fast as normal hemlock. The tips sweep upward and outward. The branchlets are slightly crowded. It is the oldest tree in the LITTLE-LEAF GROUP recorded by the Author. A more satisfactory name should be found.

- - **'Parvifolia'** Veitch, Man. Conif. 115. 1881 (*Abies canadensis parvifolia*). LITTLE-LEAF GROUP. A bushy plant, usually with multiple leaders which spread gracefully outward and upward leaving a more or less open center; branchlets thinner than normal; leaves two-ranked, broad for their length, linear, obtuse at the apex, 3-8 mm long and medium green. The Marquess of Headfort, Kells, County Meath, Ireland reported a plant (Chittenden, Conif. in Cult. 564. 1932) that at 17 years was 9 feet high and 8 feet 6 inches wide. The plant at Far Country had been obtained from Hillier and Son, Winchester, England, in the early 1930's.

In 1938 it was not quite 2 feet high and 2 feet wide. In 1945 it was 11½ feet high and 8 feet wide. This plant has since disappeared but as I understand that it is being maintained in the Hillier collection, it seems reasonable to identify it with Veitch's plant.

- - **'Parvula'** Victorin and Rousseau, Contrib. Inst. Bot. Univ. Montreal. **36**: 13. 1940. I have been unable to trace this in cultivation.

- - f. **pendula** See p. 73.

- - **'Pendula'** Hort, ex Beissn. Handbuch Nadel. 403. 1891; Bean, Trees and Shrubs, Vol II, 606. 1914. PENDULA GROUP. Bessner published his name without description in 1887, and since it is difficult to be sure of what he had in mind from his description four years later, it is consistent with the general plan of this book to follow Welch (*Manual of Dwarf*

T. canadensis 'Matthews'. Raraflora, Feasterville, Pennsylvania, in 1972. J.C.S. No. 1210.

T. canadensis 'Mansfield'. Arnold Arboretum, Jamaica Plain, Massachusetts, in 1970.

T. canadensis 'Minuta'. U.S. National Arboretum, Washington, D.C. Accession No. 20236.

Labeled *T. canadensis* 'Minima'. U.S. National Arboretum, Washington, D.C., in 1970. Accession No. 21126.

T. canadensis 'Narragansett'. Arnold Arboretum. Jamaica Plain, Massachusetts. Accession No. 80-84.

T. canadensis 'Muttontown'. Christy Park, Nassau County, Long Island, New York, in 1973. J.C.S. No. 1803.

T. canadensis 'Pendula'. A typical specimen of the European form with a central trunk.

T. canadensis 'Outpost', in the former Outpost Nurseries, Ridgefield, Connecticut. The original plant in 1945.

Conifers 1979) and restrict the use of this epithet as a cultivar name to the first clear international description we have of it, given to us by Bean. "A very attractive shrub or small tree forming a hemi-spherical mass of pendulous branches, completely hiding the interior". This very aptly describes the form found in European culture. This form does not, in contrast to all forms of Sargent hemlock, form a main framework of massive, ascending branches supporting a thatch of foliage, but maintains a vertical leader (not without a little human encouragement, I daresay, in its early days) and a framework of quite slender branches which ramify into numerous pendulous branchlets which (as Bean aptly states) produce a hemispherical mass of dense foliage that completely hides the interior. It gains height, not by extension of a few main branches, but by the death of branches smothered by later, over-lying growth.

- - **'Pendula Argentea'** Masters, Handl. Conif. Kew ed. 2: 71. 1903. Tree: branches pendent or slightly pendulous, the ends whitish. Discovered in Kew Gardens and probably no longer in cultivation.

- - 'Picta Alba' See 'Albo-spica'.

- - 'Pigtail' See 'Horsford Contorted'.

- - 'Pincushion' A listed name. I have no information.

- - 'Plainview' Another listed name on which I have no information.

- - 'Pomfret' Swartley ex den Ouden and Boom, Man. Conif. 462. 1965. A later, illegitimate synonym of 'Hiti.'

- - **'Popeleski'** New cultivar. CONICAL GROUP. Attractive, irregular, compact dwarf with moderately crowded leaves and branchlets. An early propagation was observed in 1970 at the nursery of Joseph Cesarini at Sayville, Long Island, New York. At that time it was 15 inches high and 24 inches wide. Mr. Cesarini obtained this variant from Joseph Popeleski, Greenlawn, Long Island, New York, where the mother plant is distinguished as Mk. 2.

- - f. **prostrata** See p. 73.

- - **'Prostrata'** Bean, Trees and Shrubs Vol. 3, 482. 1933. W. J. Bean, Curator at the Royal Botanic Gardens at Kew, England saw an apparently quite prostrate variety in the garden of a Mr. Renton, Branklyn Park, Perth, Scotland in June, 1931 which was clearly not the 'Pendula' he knew in European gardens. It is now (1983) a tree 20 feet across (growing on a steep bank) and is without doubt a "Sargent hemlock"—possibly of the low-growing clone 'Brookline'. This name must therefore be discarded under Art. 42 of the Cultivated Code because "the cultivar of which it purports to be the name neither does nor did exist." But the name, having been "used", must never again be re-used as a cultivar name.

- - **'Pumila'** Ordnung ex Beissn. Handbuch Nadel. ed. 2, 89. 1909. CONICAL GROUP. A regular, very compact, broad-conical form about as broad as high and with crowded branchlets and a single leader, raised by Ordnung at Eisenberg in Czechoslovakia. In 1939 the author observed a plant answering to this description in the Durand Eastman Park in Rochester, New York. It was labelled *T. canadensis pumila*, the origin not being on record. The leaves are closely set and quite divergent, distinctive through the branchlets lying in diverse planes. It seems reasonable to identify this with Beissner's plant. In 1939 the plant was 3 feet 9 inches high and 4 feet across. A grafted plant about eighteen years old was about 6 feet high in 1973.

- - 'Pygmaea' Hort. See 'Abbott's Pygmy'.

- - f. **pyramidalis** See p. 73.

- - 'Pyramidalis' Vermuelen Nurs. Cat. 1942-1959. See 'Vermuelen's Pyramid'.

- - 'Raraflora Procumbent' Provisional name. Listed Raraflora Nurs. Cat. 1970 (Name only.) Relatively fast growing, many leaves and twigs are twisted. Twigs are markedly pubescent. Leaves long, to 20 mm. This was grown from a cutting taken from a plant in the wild by Fred Bergman, Feasterville, Pennsylvania. It measured 10 inches high and about 4 feet wide. A more acceptable name should be registered if this clone merits retention.

- - 'Raraflora Pygmy' See 'Muttontown'
- - 'Raraflora Snowflake' See 'Snowflake'.
- - 'Recurva' Hort. See 'Verkade's Recurved'.
- - **'Redding'** Stout, New York Bot. Gard. Journ. **40**: 233. 1939. DENSE-LEAF GROUP. This is named from a tree standing close to Highway 107 near its junction with Route 53 near Redding, between Danbury and Georgetown, Connecticut. In 1938, when seen by the author, it was 45 feet high and the trunk measured 6 feet 7 inches in circumference. It is remarkable because of the extreme denseness of the foliage. It was discovered and introduced into cultivation by the then Outpost Nurseries (now the J. M. Woodcock Nursery) of Ridgefield, Connecticut. Dr. Stout records seeing a number of seedlings nearby, only a selection of which were like the mother plant, so this is evidently one variant not coming true from seed.
- - **'Rhapsody'** Provisional name. An interesting dwarf of the Hussii type with a congested, irregular growth pattern; leaves ¼ inch long by ⅛ inch shorter than on typical hemlock, dark green, held in double rank. The annual growth is mostly ½ inch or less, but occasional shoots develop to 2 inches. According to Dirk van Heiningen, proprietor of the van Heiningen Nurseries, Dover, Pennsylvania, about eleven plants were grown from cuttings taken from a seedling variant in the South Wilton Nurseries, Wilton, Connecticut. These plants are about 12-18 inches high and fifteen to eighteen years old. The mother plant no longer exists, nor is there any record of its appearance.
- - 'Rhode Island' Hort. or 'Rhode Island University' Hort. See 'Stewart's Gem'.
- - **'Rock Creek'** New name. GLOBOSE GROUP. A compact globe with three trunks and flexible fanlike branches, the laterals outgrowing the terminals; needles dark green, openly spaced at the growing tips. This clone, which maintains its shape better than most, originated from a collected seedling and was introduced by the Rock Creek Nurseries, Rockville, Maryland. It was at least forty years old and was 5 feet 6 inches high and 6 feet wide in 1970. The clonotype is Accession 20948 at the U.S. National Arboretum, Washington, D.C., where it was formerly grown under the name 'Globosa'.
- - **'Rockland'** Registered name. Wyman, Arnoldia **23**: 92. 1963. DENSE-LEAF GROUP. This originated in a seedling discovered in 1952 when it was about fourteen years old. It was registered at the Arnold Arboretum, Herman Brandt being recorded as the discoverer and Robert W. Pugh of Spring Valley, New York as the introducer. His description is "A vigorous compact growing hemlock whose rapid speed of growth is comparable with the growth of the species and which has a deeper green color throughout the whole growing season." Wyman adds "It also has a habit of developing numerous branchlets on the sides of current growth with a resultant heavier, denser and very compact type of growth, with approximately 2 to 3 times as many leaves per unit of stem as in the species."

I consider it unlikely that the leaves are this close; I have never observed the occurrence of even twice the normal density.
- - **'Rockport'** Den Ouden and Boom, Man. Conif. 452. 1965. (Heistad, *Florist and Nursery Exchange,* Aug. 30, 1930. *Compacta nana)* CONICAL GROUP. A broadly conical dwarf with rounded top; branchlets closely set and very slow in growth, leaves small, foliage very dark green. It was found around 1925 by H. Heistad of Rockport, Maine, on Haystack Mountain, about ¼ mile from the village of Liberty, and subsequently transplanted to his home on Amsbury Hill, Rockport.

The tree, which measured 8 feet high and 8 feet 6 inches across, was carefully lifted and burlapped, the total weight being 2300 lbs. Mr. Heistad wrote to the Arnold Arboretum and Mr. E. H. Wilson later sent a representative to examine the tree, who likened it to one discovered in Andover, Massachusetts in 1882 and then growing in the Arboretum, the Rockport tree being much the better of the two. Mr. Heistad passed away in 1945, but Mrs. Kononen, his daughter, still lives in the house and writes (15 Nov. 1982) that the plant is alive and well. It has about doubled in size in 60 years, has never set cones and is still a very dense broadly conical tree, having never been pruned

T. canadensis 'Pigtail'.

T. canadensis 'Pomfret'. Far Country, Philadelphia, Pennsylvania, in 1945. J.C.S. No. 1219.

T. canadensis 'Popeleski'.

T. canadensis 'Prostrata' Bean, at 'Branklyn' Perth, Scotland.

T. canadensis 'Raraflora Procumbent'. Raraflora, Feasterville, Pennsylvania.

T. canadensis 'Rock Creek'. U.S. National Arboretum, Washington, D.C., in 1970. J.C.S. No. 1790.

T. canadensis 'Rugg's Washington Dwarf'. R.H.S. Wesley Gardens, Woking, Surrey, U.K.

T. canadensis 'Sargentii'. A staked plant at Raraflora, Feasterville, Pennsylvania in 1970.

or propagated.

This last fact requires us to treat the published name (*nana compacta*) as invalid, since a single plant cannot claim cultivar status. It would be a pity to lose sight of this plant which appears to be attractive—and certainly is slow-growing—and in view of plans being made at last to distribute material from the tree, I propose the name 'Rockport' as a cultivar name to record the present home of the clone-type.

- - 'Rugg's Dwarf' Hort. See 'Rugg's Washington Dwarf'.

- - **'Rugg's Washington Dwarf'** Warner ex Hillier, Dwarf Conifers 66. 1964. CINNAMON-TIP GROUP. An irregular dwarf plant, resembling a heath. A plant in U.S. National Arboretum, Washington, D.C., from the Gotelli Arboretum, 1962, (Accession No. 24587) obtained from Ralph M. Warner of Milford, Connecticut in 1957, was in 1983 6½ feet high and 8 feet wide. It is of an upright, loose, globose tree. Current and second year shoots are cinnamon colored. Medium green needles are sparsely spaced on the young shoots.

- - 'Ruggii' Hort. or 'Ruggsii' Hort. See 'Rugg's Washington Dwarf'.

- - 'Saint George's Dwarf' A listed name. Not sufficiently distinct from 'Hussii' to justify retention.

- - **'Salicifolia'** Verkade, Nurs. Cat. LITTLE-LEAF GROUP. This is named from a small tree on the grounds of the Detmer Nurseries, Tarrytown, New York. It was imported from Germany more than sixty years ago. It has been propagated by Verkade's Nurseries, Wayne, New Jersey, and sold as 'Salicifolia'. This is a more compact plant than 'Parson's Little-Leaf', with leaves 4-8 mm long, somewhat crowded and a darker green color.

- - 'Saratoga Broom' Hort. See 'Beaujean'.

- - 'Sargent Broom' A listed name. It seems insufficiently distinct from 'Cole's Prostrate' to merit retention.

- - **'Sargentii'** Scott F. J.', *Art of Beautifying Home Grounds* 549. 1870 (*sargenti*); Sargent H. W. in Downing, A. J., *Landscape Gardening* 581. 1875. ("pendula" pro. syn.); Parsons Nurs. Cat. 59. 1879; var. *sargentiana* Kent in Veitch, Man. Conif. 465. 1900; var. *Sargentii* Bean, Trees and Shrubs 606. 1914. PENDULA GROUP.

THE SARGENT HEMLOCKS

This cultivar (or group of cultivars, as I here suggest it should now be regarded), is probably the most important (certainly the best known) cultivated variation of Canadian hemlock. It owes its importance to two facts. (1) It was the first to be selected in American horticulture and so has had time to develop in protected surroundings into magnificent specimens, and (2) an enterprising nurseryman in New York soon "spotted" the plant, realised its commercial potential and set about introducing it into cultivation, as a result of which many daughter plants exist, themselves now of considerable maturity. At first (1857) it was described as "seedlings", later (1868) as "dwarfish" and later still (1870) as "a flat-headed tree or a great bush", so no doubt some of the new forms now being planted in gardens have the same potential and in 100 years' time will rank with some of the finest Sargent hemlocks of today.

The history of this complex seems to be shrouded with a mystery and glamour that is quite unnecessary. As early as 1803 Michaux had recorded this abnormality (see p. 63) but almost certainly the first direct reference to be found is in the first American book on conifers—Hoopes (*Book of Evergreens*) published in 1862. There (p. 418) in describing his visit to the arboretum of Henry Winthrop Sargent at Fishkill, New York, he writes: "We noticed a remarkable variety of the Hemlock Spruce, of dwarfish habit, with long, drooping branchlets, and altogether quite unique in character".

Mr. Charles Jenkins, (*Hemlock Arboretum Bulletin* 4 July 1933) gave a historical account that is substantially accurate. Alfred J. Fordham, in an article *Four Fathers of the Sargent Hemlock* (*American Nurseryman* 96(12): 9. 1962) brings this account up to date, and recently Peter del Tredici, propagator at the Arnold Arboretum, has been rummaging in the old books and has turned up several early references. His paper *Sargent's Weeping Hemlock Reconsidered* (*Arnoldia* **40**(3): 202. 1980) is well worth studying by anyone interested to go into more detail than can be found space for in this book. He comes up with some interesting discoveries; one concerns the date of the finding of the plants. Exact dating seems to be impossible, since all we have are memories forty years or so old. Earlier estimations range between the years 1857 and 1870. Del Tredici feels able to narrow it down to the years 1857 and 1868. Allowing time for the young transplants to settle down and attract attention and also (this as a nurseryman myself I can fully understand) the years that must elapse between acquiring a new variety and building up stock of plants of saleable size, for it to have figured in a nursery catalogue of 1879, the date of finding must have been at the beginning of that period—perhaps 1857, or at most "the late 1850's".*

The earlier references now drawn to our attention all confirm that from the very first the plants were regarded as selections, i.e. they were cultivars. This justifies the adoption here of the treatment first proposed by Welch (*Dwarf Conifers* 323. 1966) of listing all these pendulous forms within a PENDULA GROUP.

But how many clones are we dealing with? Leaving aside the fact that pendulous specimens of *Tsuga canadensis* can (and do) arise in the seed-beds from time to time and the fact that a fifth clone (see 'Horton') of comparable age and having an undefined but almost certain connection with the original four Sargent hemlocks is now known, what of these latter?

There were four plants and they were distributed by the supposed finder, General Joseph Howland, as follows:

Tree No. 1   Kept by the General at his own house, "Tioronda", Matteawan, New York. (Now the Craig House Sanitarium, Beacon).

Tree No. 2   Given to his neighbour and cousin, Henry Winthrop Sargent, "Wodenethe", Fishkill-on-Hudson (now also in the township of Beacon, New York) and named in his honour.

Tree No. 3   Given to H. H. Hunnewell, Wellesley Arboretum, Massachusetts.

Tree No. 4   Given to Professor Charles S. Sargent, "Holmlea", Brookline, Massachusetts. He was Director of the Arnold Arboretum.

Trees Nos. 1 and 4 are still thriving. Hornibrook (*Dwarf Conifers* 185. 1923) stated that Trees Nos. 2 and 3 were dead, giving no authority for the statement, and this seems to have been uncritically accepted by all subsequent writers. But during a recent visit to the Hunnewell Arboretum, Dr. T. R. Dudley was informed by Mr. Willard Hunnewell (the present owner) that the records are so sketchy that it would be impossible to say whether or not the huge Sargent hemlock now growing there is the original gift from General Howland or one of the Parsons' propagations. Judging from the size of the plant and its unusual branch-structure, the former supposition is reasonable. See p. 115.

---

*Since the above and what appears on the subject in Chapter Two of this book were prepared for press, Peter del Tredici's book *A Giant among the Dwarfs* has appeared. It is a very useful contribution to the study of the Sargent hemlocks, the early appearances of the name in the literature being set out in full. He has come up with a totally new theory regarding the origin of the Sargent hemlocks, viz. that there were five seedlings collected by the "original" Mr. Horton, one of which he kept (and being a practical gardener, stem-trained): the four others being transferred, probably to H. W. Sargent, and distributed by him in the manner generally accepted. This is a most plausible explanation, although it is not free from the weakness of the older account—that it ultimately rests on conjecture. It would be considerably strengthened if authentic propagands of the 'Horton' clone developed the trunkless, multi-branched habit of the other Sargents.

Del Tredici gives a scientific account of the growth pattern of the Canadian hemlock and some explanation of the variation in this pattern in the pendulous forms that does not (I am pleased to note) discredit my own simple gardeners' explanation on pp. 45-46. Both accounts explain "How?" Neither answers the question "Why?"

Welch (*Dwarf Conifers* 322. 1966) reports being told that the tree at the Arnold Arboretum can be traced to Tree No. 1, which was "the only one to have been widely propagated, the Parsons Nursery having sent a propagator up to Matteawan for grafts." This would have been Tree No. 1, but their 1879 catalogue informs us that they had received it from H. W. Sargent of Fishkill-on-Hudson, the owner of Tree No. 2. So it is quite possible for all four clones to be still extant. This possibility throws some doubt on the clonal identity of the tree in the Arnold Arboretum selected by Welch (*Manual of Dwarf Conifers,* 376. 1979) as the clonotype: it would be safer and more logical to cite Tree No. 1 (at Matteawan) for this honour, since that tree is still available.

Article 2 of the *Cultivated Code* lays down that a further selection from within a cultivar (itself being, by definition, a selection) is to be regarded as a distinct cultivar. Welch (*Dwarf Conifers,* 323. 1965) first gave effect to this principle in the present case, by suggesting the name 'Brookline' for the clone deriving from Tree No. 4. With the probability now suggested that living material of Trees Nos. 2 and 3 is still with us, I now propose a logical completion of this by naming these clones respectively 'Wodenethe' and 'Hunnewell'. ('Wellesley' would be too easily confused with 'Wellesliana'. See p. 84). Finally 'Horton' offers itself as an obvious choice for the tree found in the town of that name.

'Sargentii' as a cultivar name is therefore here used as the strictly correct epithet for the single clone derived from Tree No. 1. Each of the four (or, if you please, five) clones is now now identifiable as a distinct cultivar, although doubtless the term "The Sargent hemlocks" will continue in use in a collective sense. For this reason the cultivar name 'Tioronda' suggested by del Tredici (*A Giant Among The Dwarfs* 40. 1983) has much to commend it. Rules should be our servants, not our masters.

All forms of Sargent hemlock form large shrubs (this is true despite their tree-like size) semi-globose, broader than high (in the case of 'Brookline' much broader) with no vertical trunk (unless artificially produced by early training in the nursery) but with a framework of a few massive, ascending main branches (often heavily deepened for strength) and an irregular thatch of foliage carried on pendulous terminal shoots.

It remains for some means to be found for identifying and distinguishing the clones. Habit and tree shape here is useless—it depends too much upon early cultural practices. Leaf and spray characteristic may or may not be adequate and probably the only sure method would be microscopic examination of the foliage, similar to work that was done in Britain to disentangle the different clones of the Leyland Cypress or gas chromatography or some other resource of present-day technology. What about it, Peter?

- - **'Schramm'** New cultivar. 'Schramm Fastigiate'. Hort. FASTIGIATE GROUP. A narrowly conical form with the branches on the lower third of the trunk strikingly fastigiate; foliage and growth rate about normal. It originated as one (the larger) of two similar trees found by Dr. J. R. Schramm, former director of the Morris Arboretum about one mile southeast of Compass, Pennsylvania, between Downingtown and West Chester, on Route 122 in a farm and described by him in a letter to Mr. Jenkins. In 1972, the mother tree was 8 feet 8 inches in circumference at breast height and about 75 feet in height. According to the present owner of the farm these trees were planted about 1890.

- - **'Sherwood Compact'** New cultivar. SPREADING GROUP. A seedling raised at the Sherwood Nursery, Portland, Oregon. It forms a low-growing mound with ascending, somewhat twisted branches, giving a very attractive appearance to the dark green leaves.

- - **'Silver Tip'** Helene Bergman in Plants and Gardens 21(1): 42. 1965. WHITE-TIP GROUP. A selected seedling at Raraflora, Feasterville, Pennsylvania, relatively free-growing, with the white-tipped branches shorter than usual. It has ascending branches with pendulous tips.

- - **'Silvery Gold'** New name. ('Bergman's Aurescens Nana' Raraflora Nurs. Cat. c. 1970. Name only.) GOLDEN GROUP. Similar to 'Everitt Golden' but quite silvery in extreme winters. The illegitimate name has had to be replaced.

- - **'Slenderella'** Hort. New cultivar. LITTLE-LEAF GROUP. This has small, light-green leaves, but unlike many other little-leaf clones it has an open and rapid growth. John Mitsch writes that it is difficult to propagate from cuttings. He obtained it from Longwood Gardens, Kennett Square, Pennsylvania. It was at one time distributed under the name 'Julians'.

- - **'Snowflake'** New name ('Raraflora Snowflake' Hort.) WHITE-TIP GROUP. This is another free-growing bushy plant showing good white color. It was grown from a cutting taken from a bud sport in 1959. In 1970 this plant was 28 inches high and 42 inches wide. In 1972, it was 4 feet 6 inches high and 7 feet wide. Two wet summers in succession have increased its growth rate. The simpler name is proposed to avoid confusion. 'Snowflake' is difficult to distinguish from 'Albo-spica.'

- - f. **sparsifolia.** See p. 73.

- - **'Sparsifolia'** Beissn. Nadel. 402. 1891. SPARSE-LEAF GROUP. Compact shrub, branches upright; foliage with a sparse or lacy character due to the irregular arrangement of the leaves. This cultivar is probably no longer in cultivation.

- - 'Spingarn's Prostrate Weeper' Hort. Provisional name. This I have not been able to trace in cultivation, unless it is the same as 'Baldwin Prostrate' listed Spingarn Nurs. Cat. 1970 (Name only.) If it is still in circulation and worth retaining, a more acceptable name should be found and registered.

- - 'Sport of Coles' Hort. See 'Helene Bergman'.

- - 'Sport of Jervis' Hort. See 'Jervis'.

- - **'Starker'** New name. 'Minuta Pendula' Helene Bergman, Plants and Gardens **21**(1): 44. 1965; Spingarn, Amer. Rock Gard. Soc. Bull. **27**(3): 87. 1969; 'West Coast Spreader' Spingarn, loc. cit.; 'West Coast Creeper' Spingarn, loc. cit. PENDULA GROUP. This is presumably a seedling of 'Sargentii' but it is rather slower in growth. It makes a plant more than twice as broad as high. A plant at Raraflora, Feasterville, Pennsylvania produces a small percentage of bearded cones. These have leaf-like appendages that appear between the cone scales. This cultivar originated with Carl Starker of Jennings Lodge, Oregon. He called it *T. canadensis minutae pendula* in letters to Mr. Bergman prior to 1959, but a new name has been assigned to avoid confusion.

- - 'Stewartii' Hort. See 'Stewart's Gem'.

- - **'Stewart's Gem'** New cultivar. CINNAMON-TIP GROUP. This cultivar has been propagated by Robert S. Stewart, owner of Stewart's Nursery, Wakefield, Rhode Island. According to Mr. Stewart, his father found a small plant in 1936 at Quinebaugh on the border of Vermont and New Hampshire. A cutting-grown plant about twenty-eight years old was 15 inches high and 3 feet wide. It usually does not grow this wide. 'Stewart's Gem' is more compact than most clones of this group. Lloyd Lawton, Tiverton Four Corners, Rhode Island, lays claim to a slower-growing plant which he had obtained from Mr. Stewart, Sr. many years ago. The author observed this plant in 1972 and judged it to be not nearly as old as the sixty-five years Mr. Lawton claimed, and probably the same clone. 'Stewart's Gem' is represented at the Arnold Arboretum by Accession No. 1088-65, obtained from Stewart's Nursery by Professor J. Caddock, University of Rhode Island, Kingston, Rhode Island. This cultivar is also known by the commercial synonyms 'Rhode Island' and 'Rhode Island University'.

- - 'Stewart's Pygmy' Hort. See 'Stewart's Gem'.

- - **'Stockman's Compact'** Vermuelen ex Lewis, Amer. Nurs. **114**(4): 45. 1961. A slow-growing but not in the end a dwarf form. Selected as a seedling in Stockman's Nursery, Greenlawn, New York, near West Stockbridge, Massachusetts, and introduced to the trade by Vermuelen Nursery, Neshanic Station, New Jersey (Nurs. Cat. 1960.) Previously, from 1941, it had been distributed as 'Compacta MK 2'.

- - **'Stockman's Dwarf'** Vermeulen ex Lewis, Amer. Nurs. **114**(4): 45. 1961. Grows with tight mounds of prune-type growth. A very attractive form. Introduced to the trade by Vermeulen Nursery, Neshanic Station, New Jersey. (Nurs. Cat. 1960.) Previously, from

T. *canadensis* 'Sargentii' trunk.

T. canadensis 'Starker' at Kingsville.

T. canadensis 'Schramm'. Compass, Pennsylvania, in 1945. J.C.S. No. 1613.

The same tree: close-up of fastigiate branch system.

T. canadensis 'Stockman's Dwarf'.

T. canadensis 'Stranger'. Far Country, Philadelphia, Pennsylvania, in 1945. J.C.S. No. 1456.

T. canadensis 'Taxifolia'. Arnold Arboretum, Jamaica Plain, Massachusetts, in 1970. J.C.S. No. 1369.

T. canadensis 'Towson'. Towson Nurseries, Towson, Maryland, in 1938. J.C.S. No. 1432.

1941, it had been distributed as 'Compacta MK 1'.

- - **'Stranger'** Swartley ex den Ouden and Boom 463. 1965; 'Strangeri' Jenkins, Heml. Arb. Bull. 4. 1933. This clone forms a compact tree about as broad as high, leaves are variable in size reaching to 12 mm × 2 mm, tapering to a blunt or rounded tip, held pectinately at 80°-90° or (2nd year shoots) at random angles. The color is a deep rich green above, with two wide stomatic bands beneath. The original tree in the Cherry Hill Nurseries, Newbury, Massachusetts, was 12 feet high in 1938 and it still (1970) displays an unusually broad top. A grafted plant in Far Country was 2 feet 3 inches high and 2 feet wide in 1938. In 1945 it was 6 feet 6 inches high and 5 feet 3 inches wide.

- - 'Summer Snow' conical form resembles 'Alba spica' but midway it and 'Gentsch White'. Pure white new growth at tips. 15 years 6' high and 4½' wide. Unknown seedling propagated by John Mitsh, Aurora, Oregon.

- - 'Tannersville Broom' Provisional name. Listed Hillside Nurs. Cat. c. 1970. (Name only.) I have not seen it.

- - **'Taxifolia'** Abbott ex Jenkins, Heml. Arb. Bull. 11. 1935. YEW-LIKE GROUP. A slow-growing plant with yew-like foliage: leaves very irregular in arrangement, much longer than typical, many curved and narrowing to the apex, which is obtuse to acutish. A grafted plant at Far Country, about as broad as high in 1938, was slightly more than 1 foot high and in 1945 was nearly 3 feet high. This plant has since disappeared but plants have been observed in other collections.

- - **'Thurlow'** New cultivar. This is a fast-growing clone of the 'Hussii' type that originated at Cherry Hill nursery, Massachusetts where it was noticed by Layne Ziegenfuss and Gregory Williams who later introduced it to the trade. It is compact when small, but opens up with age. The leaves are very small, ⅜ inches × ⅛ inches, dark green, held in double rank. The main branches are held at 60° to the trunk and are covered with foliage. It was named in honour of Mr. Thurlow, part owner of Cherryhill Nursery.

- - **'Towson'** Swartley ex Helene Bergman, Plants and Gardens **21**: 43. 1965, den Ouden and Boom, 464. 1965. LARGE-LEAF GROUP. Although this is the same cultivar as that described by den Ouden and Boom from Towson Nurs. Cat. 1931, they incorrectly state that the leaves are short whereas actually the leaves are 19 mm long, clearly placing it in the LARGE-LEAF GROUP. It is a slow-growing compact tree with narrow habit when young, with long and densely set leaves, not in two ranks, very dark green; branches stiffly descending, their tips only slightly recurving. Branchlets short and very crowded. Leaves long and broad, densely set. Den Ouden and Boom were incorrect in stating that it is no longer in cultivation. It was propagated by grafting at Rock Creek Nursery, Rockville, Maryland. Mrs. Bergman gave the size of the same plant as 8 feet high and 4 feet wide. As so often happens, the relative height and spread may change as an individual plant grows to maturity.

- - 'Tucker's Selection' Provisional name. Listed Vermeulen, 1941. I have not seen it. A rather more imaginative name should be found if the clone is sufficiently distinctive to be worth retaining.

- - **'Unique'** New cultivar. GLOBOSE GROUP. A dense, bun-type plant selected as a seedling in the South Wilton Nurseries, Wilton, Connecticut. The branchlets are crowded, the leaves are ⅜ inches × ⅛ inches held in double rank. After 28 years the plant in 1983 is 8 feet high and 7 feet wide.

- - **'Upper Bank'** New cultivar. This originated as a tree on the Upper Bank Nursery, Media, Pa. which was seen by Layne Ziegenfuss and Gregory Williams, who later introduced it to the trade. It is quite slow-growing, forming a conical plant of neat outline. The leaves are ¼ inches long by 1/16 inches wide, and are light green. At 14 years of age a plant is 14 inches wide by 26 inches high.

- - **'Valentine'** New name. ('Valentine's Weeping' McDonald, Univ. Tenn. Arb. Soc. Bull. 3: 1967.) PENDULA GROUP. This was collected as a wild seedling in 1941 by the Valentine Nursery, Cosby, Tennessee. In February 1966 it was moved to the University of Tennessee Arboretum, when it was described as follows: "It has three main branches which form a 'Y' with a spread of 18 feet at the widest point. The plant is 4 feet 3 in. high in the center but the main trunk is only 13 in. high and had a diameter of 5 in. . . . The branch ends recurve near the ground." This was at the age of twenty-six years. It has been distributed to a number of arboreta and nurseries.

- - 'Valentine Yew-like' Provisional name. YEW-LIKE GROUP. This plant was sent to Far Country by the Valentine Nurseries, Cosby, Tennessee, in April 1939 bearing the label *T. canadensis taxifolia.* Mr. Jenkins gave this plant to Swarthmore College, Swarthmore, Pennsylvania, about 1945, under this name. In 1972 it was 45 feet high and 13 feet wide. This is quite an attractive and graceful tree and is being propagated.

Recently the foliage of 'Jennings' and 'Valentine Yew-like' have undergone some changes. Some parts of the plant are going back part way to typical hemlock. This is not a matter of reversion on certain branches: it is not that clear cut. If this clone is worth retaining a name should be registered that will not cause confusion.

'Taxifolia' is probably the most valuable clone in this group from a collector's standpoint.

- - **'Van Dyne'** New cultivar. CINNAMON-TIP GROUP. The leaves of this cultivar are nearly of normal length, mostly pointed, some very narrow; the twigs have no swelling at the tips, and the hairs are shorter than in 'Cinnamomea'.

- - **'Variegata'** Jäger u Beissn. Ziergeholze; 445. 1884. (*Abies canadensis variegata*) Merely described as "with yellowish-white leaves." The clone can no longer be identified in cultivation, so this name should not be used.

- - 'Variegata Gentsch' Hort. See 'Gentsch White'.

- - **'Verkade Petite'** Registered name. Verkade ex Wyman, Arnoldia **29**: 8. 1969. John Verkade of Verkades Nurseries, Wayne, New Jersey, supplied the following information: "This little gem was found as a seedling (a three-year-old plant in 1955). It has an annual growth of 1/6 to 1/8 inch. It has a cluster of tiny brown eyelets on each of its tiny branches and in the spring it is a mass of light green when the new growth takes place. 'Verkade Petite' must be planted in the shade as it cannot adjust to the summer sun, but it is very hardy and will not be harmed in minus ten degrees. The parent plant is now (1968) sixteen years old, 2 in high and 3½ in across, and is an irregular globe."

This interesting plant is one of the dwarfest of hemlocks; it is certainly in the class of 'Abbott's Pygmy', but due to the fact that it is not as healthy and rugged a plant as that cultivar it suffers a certain amount of die-back and so it is difficult to gauge its true rate of growth.

- - **'Verkade Recurved'** Registered name. Verkade ex Wyman, Arnoldia 29: 8. 1969. This attractive form was registered by Mr. John Verkade. He states: This clone has a very irregular growth habit, with an annual growth of approximately 2 to 3 inches per year. It was found in a hemlock collector's nursery and given to us as a misformed plant that refused to grow despite additional feeding. The parent plant, estimated to be 15 years old, is 16 inches tall, and 12 inches wide with a pyramidal habit is now in our gardens in Pompton Lake, New Jersey. This clone is open-growing with recurved needles which do not at all resemble those of the normal *Tsuga canadensis.*"

- - 'Verkade's Pincushion' Hort. See 'Pincushion'.

- - 'Verkade's Recurva' See 'Verkade Recurved'.

- - 'Verkade's Sparseleaf' Provisional name SPARSE-LEAF GROUP. In 1973 a small plant was observed in the collection of Joel Spingarn, Baldwin, Long Island, New York. This was a recent acquisition from Verkade's Nurseries, Wayne, New Jersey. It was 20 inches high and 16 inches wide, with sparse leaves, the foliage having a juvenile appearance.

- - 'Verkade's Witches' Broom'. See 'Kathryn Verkade'.

- - **'Vermuelen's Pyramid'** Vermuelen, Nurs. Cat. 1960. 'Pyramidalis' Vermuelen, Nurs. Cat. 1942. FASTIGIATE GROUP. Narrow, compact, and pyramidal habit with densely set and radially disposed leaves: resembles *T. caroliniana* from a distance; it originated as a seedling collected in the wild by W. C. Horsford, Charlotte, Vermont, about 1915. It grows into a narrow column with drooping branches but just misses being a striking

T. canadensis 'Verkade's Broom'.

T. canadensis 'Vermuelen's Pyramid'. Arnold Arboretum, Jamaica Plain, Massachusetts, in 1970. Accession No. 518-62.

T. canadensis 'Verkade Recurved'. Collection of Joel Spingarn, Baldwin, Long Island, New York.

T. canadensis 'Von Helms Dwarf'. Far Country, Philadelphia, Pennsylvania, in 1940. J.C.S. No. 1227.

T. canadensis 'Watnong Star'.

T. canadensis 'Warner's Globe'. Collection of Joel Spingarn, Baldwin, Long Island, New York, in 1970. J.C.S. No. 1811.

T. canadensis 'Wilton'. South Wilton Nurseries, Wilton, Connecticut, in 1945. J.C.S. No. 1783.

T. canadensis 'Youngconer'. U.S. National Arboretum, Washington, D.C., in 1970. Accession No. 20238.

plant. A plant at Arnold Arboretum (Accession No. 518-62) in 1970 measured 15 feet high and 5 feet wide.

- - **'Von Helms'** Helene Bergman, Plants and Gardens **21**(1): 42. 1965. 'Von Helms Dwarf' Swartley ex den Ouden and Boom, 464. 1965. DENSE-LEAF GROUP. Grown from seed in 1924 in the nursery of William Von Helms, Monsey, New York. Two plants were given to Far Country in 1936. The smaller one, which is the clonotype, was then 16 inches in height and the same in breadth. In 1945 this same plant measured 6 feet by 6 feet. 'Von Helms' forms a dense, somewhat irregular, broad-topped cone. Eventually it becomes somewhat open in habit. The foliage is dark green and irregularly disposed. A tree in U.S. National Arboretum, Washington, D.C. (Accession No. 20441) is 4 feet high $\times$ 5 feet wide and is similar in form to 'Brandley'.

- - 'Warner's Armistice' Hort. See 'Armistice'.

- - **'Warner's Globe'** New name. 'Warner's Globose' Hillier, Dwarf Conifers. 67. 1964. GLOBOSE GROUP. This cultivar forms an open globose bush, with strong main shoots rather sparsely clothed with leaves (15 mm $\times$ 3 mm) orange-brown, pubescent; smaller leaves occur in tufts on short spurs. Leaves are uniformly elliptic with sharp tips. The color is a rather yellowish green, above and below, the stomatic bands being quite inconspicuous. It was introduced by Ralph M. Warner of Milford, Connecticut. Specimens were observed in the U.S. National Arboretum and in the collection of Joel Spingarn.

- - **'Watnong Star'** Registered name. Don Smith ex Hebb, Arnoldia **30**: 260. 1970. WHITE-TIP GROUP. "At 11 years this plant is a soft globe 16 in. high and 20 in. across. The texture is tight but not stiff. New growth is almost white at the tips, giving the effect of hundreds of small white stars." It was found in a woodland in New Hampshire by Robert Clark of the Rochester Parks Department, Rochester, New York.

- - **'Waverly'** Jenkins, Heml. Arb. Bull. 61. 1948. DENSE-LEAF GROUP. The original plant is a large hemlock reported by Francis J. Stokes of Germantown, Philadelphia, Pennsylvania. It was found in a pasture about one mile from Clark's Green Corners on Route 407, Waverly, Pennsylvania. In 1947 this plant was about 40 feet high and 11 feet 6 inches in circumference, at breast height. Many seedlings are growing in the vicinity. These are quite variable especially in foliage characters.

- - **'Wellesliana'** Hornb. Dwarf Conif. ed. 2. 274. 1939. A low, broadly conical bush of pendulous branches, about 5 feet high by as much through. According to Hornibrook it arose as a seedling, in the Hunnewell Pinetum, Wellesley, Massachusetts, but there is no record of such a plant there. It cannot be the plant described under the proposed name 'Hunnewell' and must be regarded as lost to cultivation, even if it ever existed other than in Hornibrook's notes.

- - 'West Coast Creeper' Hort. See 'Starker'.

- - 'West Coast Spreader' Hort. See 'Starker'.

- - **'Westonigra'** Den Ouden and Boom, Man. Conif. 464. 1965. COMPACT GROUP. More compact than the ordinary hemlock; branchlets pendulous; leaves not 2-ranked, densely set, exceptionally dark green. Found as a seedling in the Weston Nurseries, Weston, Mass. about 1940.

- - **'Wheelerville'** Swartley ex den Ouden and Boom, Man. Conif. 464. 1965. CONICAL GROUP. Makes a dense pyramid of slow growth; leaves very closely set, irregularly placed, pointing in all directions. (Wheelerville is an abandoned town between Williamsport and Eaglesmere, Pennsylvania). The plant was discovered by Everett G. Logue of Williamsport, Pennsylvania. I have not been able to trace it in cultivation.

- - **'Wilton'** Swartley ex den Ouden and Boom, Man. Conif. 464. 1965. A graceful tree with a well-clothed appearance; lower branches horizontal; branchlets drooping; leaves not exactly two-ranked, glossy bright green. Raised by Jacob C. van Heiningen, South Wilton Nurseries, Wilton, Connecticut. A plant of 'Wilton' is in the collection at Ambler Campus of Temple University, Amber, Pennsylvania.

- - 'Wilton Globe' Provisional name. GLOBOSE GROUP. Foliage more or less radially disposed. Two multi-stemmed plants were sent from the South Wilton Nurseries, Wilton, Connecticut, to Far Country many years ago. If worth retaining a name should be registered that will not give rise to confusion with 'Wilton'.
- - 'Wilton Wideleaf' Provisional name. The plant near Jacob C. van Heiningen's house, Wilton, Connecticut, should be considered representative of this cultivar. If worth retention, a name should be registered that will not give rise to confusion.
- - 'Wodenethe' New cultivar. PENDULA GROUP. On p. 136 I have explained my proposal for this new name to cover the progeny of original Sargent hemlock Tree No. 2, given to Henry Winthrop Sargent (not the Professor of the same name), and planted on his estate, Wodenethe, Fishkill-on-Hudson (now included in the town of Beacon, New York).

Although the mother-tree is long since dead, this posthumous christening is not disallowed by Article of the *Cultivated Code,* since undoubtedly this cultivar (a selection from within the cultivar 'Sargentii' as originally used to cover all four of the Fishkill seedlings) DID exist, and it is justified by the following description taken from the Parsons and Sons Nursery Catalogue dated 1879. It read:

> *Abies canadensis pendula sargenti,* Sargent's weeping hemlock, the most graceful and delicately beautiful evergreen known. When the leader is trained to a stake it can be carried to any reasonable height, each tier of branches drooping gracefully to the ground, like an evergreen fountain. It was first sent out from Flushing, having been received from H. W. Sargent, of Fishkill-on-Hudson.

Dr. Stout (*Journal of the New York Botanical Garden* **40:** 158. 1939) records a letter he had received from the late Jacob C. van Heiningen. It read:

> While I worked for the Parsons & Sons Nursery Co., I heard that Mr. J. R. Trumpy, the famous propagator who was brought over on a sailing vessel by Mr. Samuel Bowne Parsons from Hugh Low & Co., London, England, before the Civil War, was sent to Fishkill-on-the-Hudson to get scions of the original plants and as far as I understand the first plants shown were at the great exhibition in Philadelphia, Pa. in 1876 and the Parsons were the first to introduce these Hemlocks in the trade.

It would therefore seem probable that, contrary to tradition, some if not all of the early propagations are of this clone and not of 'Sargentii' (i.e. in the restricted use of the epithet now suggested for the clone originating with tree No. 1.)

- - 'Youngcone' New name. Williams. A bushy hemlock that regularly bears cones when only five to six years old; foliage nearly typical, except that the lateral twigs are almost as long as the terminal. The stomatic lines are indistinct. Found many young plants, multi-stemmed and bearing cones in a pasture in Vermont. The shape varied from upright to globose, according to Greg Williams, who collected dwarf conifers assiduously for several years in company with Layne Ziegenfuss of Lehighton, Pennsylvania.

A tree at the U.S. National Arboretum (Accession No. 20238) measured 5 feet by 4 feet in 1970. In 1974 it had reached 6 feet by 6 feet. The plant is still compact with the lateral tips half weeping, with many cones of the current year plus at least one-third of last year's still hanging on.

IDENTIFICATION PLATES

Photographed by Corky Rohrbaugh

a. *Tsuga canadensis* 'Abbott's Pygmy'. **b.**--'Angustifolia'. **c.**--'Armistice'. **d.**--'Ashfield Weeper'. **e.**--'Atrovirens.'

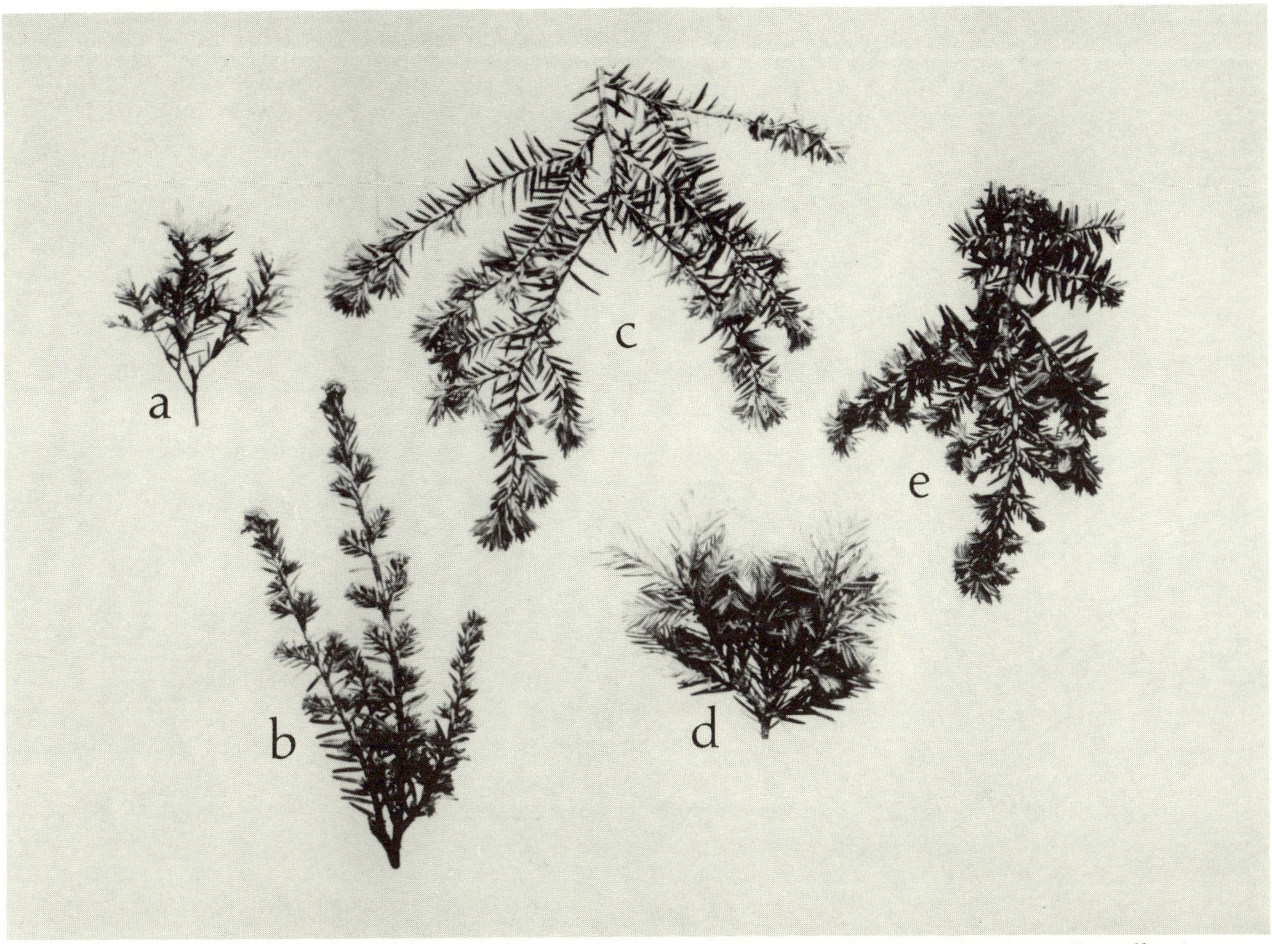

a. *Tsuga canadensis* 'Baldwin Dwarf Pyramid'. **b.**--'Beaujean'. **c.**--'Bennett'. **d.**--'Betty Rose'. **e.**-- 'Brandley.'

a. *Tsuga canadensis* 'Bristol'. **b.**--'Callicoon'. **c.**--'Cappy's Choice'. **d.**--'Coffin'. **e.**--'Cole's Prostrate.'

a. *Tsuga canadensis* 'Curtis Compact'. **b.**--'Curtis Sparse-Leaf'. **c.**--'Curtis Spreader'. **d.**--'David Verkade'. **e.**--'Densifolia.'

a. *Tsuga canadensis* 'Elm City'. **b.**--'Everitt Golden'. **c.**--'Fantana'. **d.**--'Far Country'. **e.**--'Geneva.'

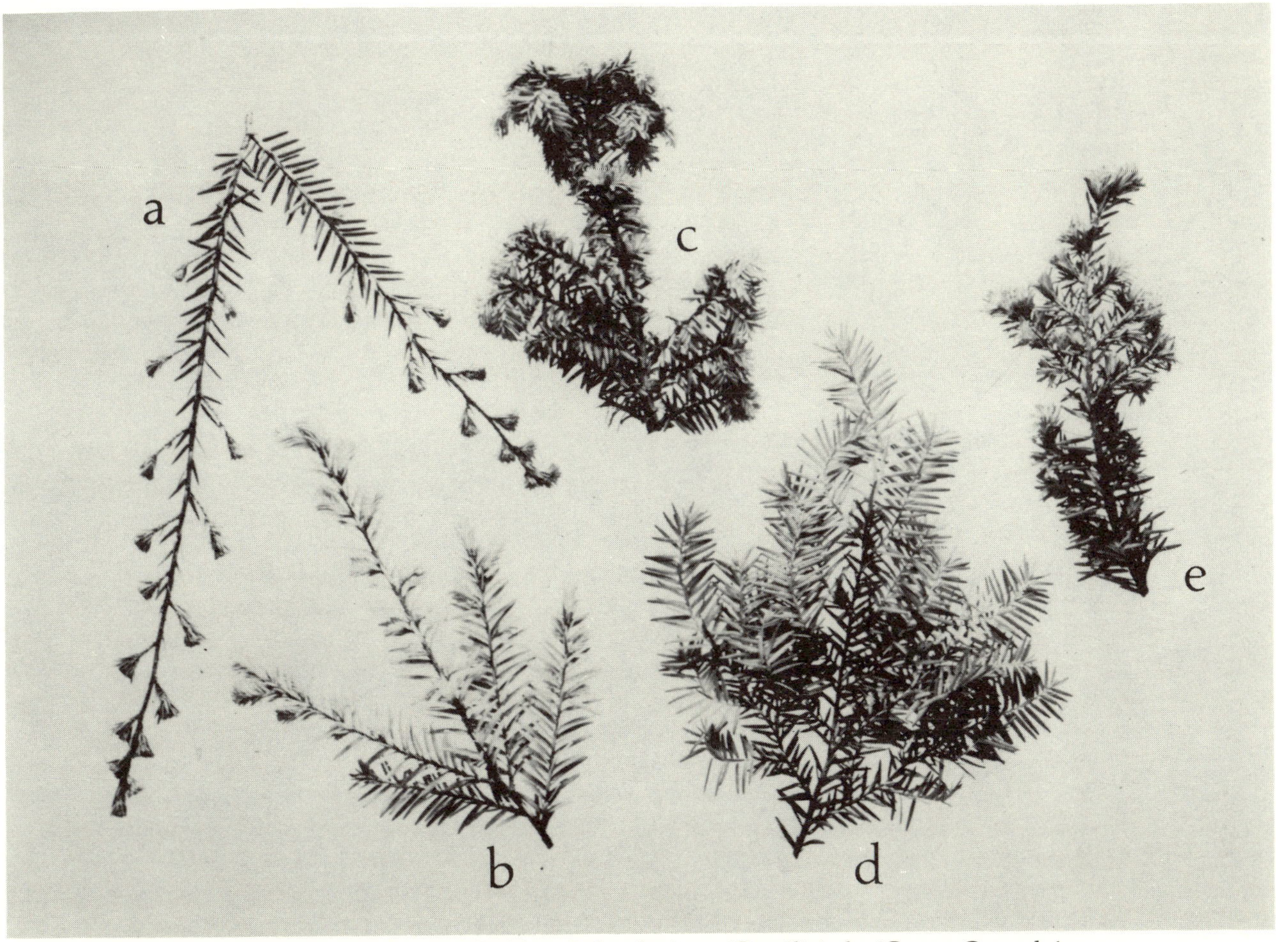

a. *Tsuga canadensis* 'Gentsch White'. b.--'Golden Splendor'. c.--'Gracilis'. d.--'Green Cascade'. e.--'Greenspray.'

a. *Tsuga canadensis* 'Greenwood Lake'. b.--'Guldemond's Dwarf'. c.--'Harmon'. d.--'Henry Hohman'. e.--'Horsford Contorted.'

a. *Tsuga canadensis* 'Jacqueline Verkade'. **b.**--'Jan Verkade'. **c.**--'Jennings Yewlike'. **d.**--'Jervis'.
e.--'Kathryn Verkade.'

a. *Tsuga canadensis* 'Kelsey's Weeping'. **b.**--'Latifolia'. **c.**--'Lewis'. **d.**--'Little Joe'. **e.**--'Lustgarten
Creeping.'

a. *Tsuga canadensis* 'Minuta'. **b.**--'Muttontown'. **c.**--'Palomino'. **d.**--'Popeleski'. **e.**--'Rugg's Washington Dwarf.'

a. *Tsuga canadensis* 'Slenderella'. **b.**--'Snowflake'. **c.**--'Stewart's Gem'. **d.**--'Stockman's Dwarf'. **e.**--'Stranger.'

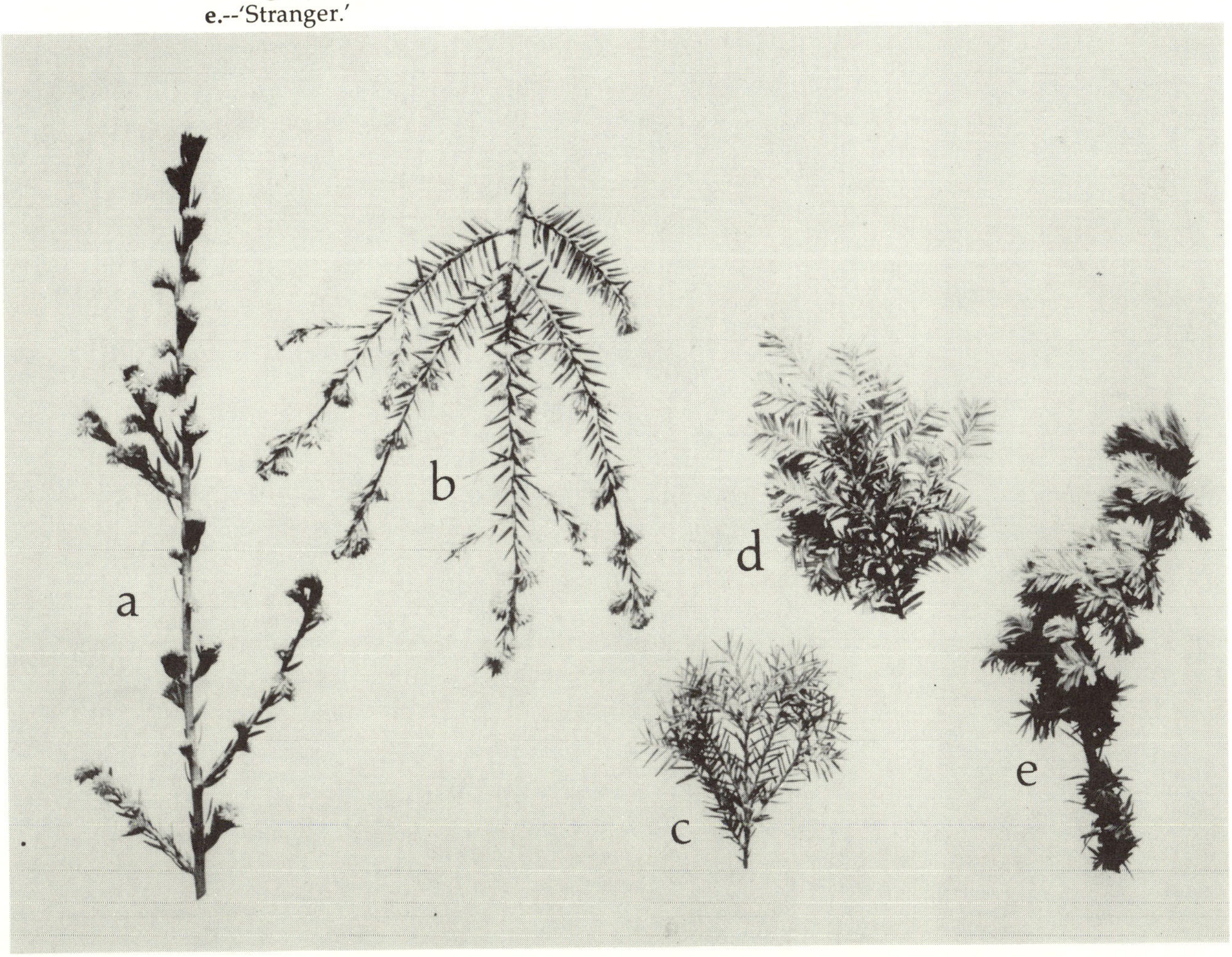

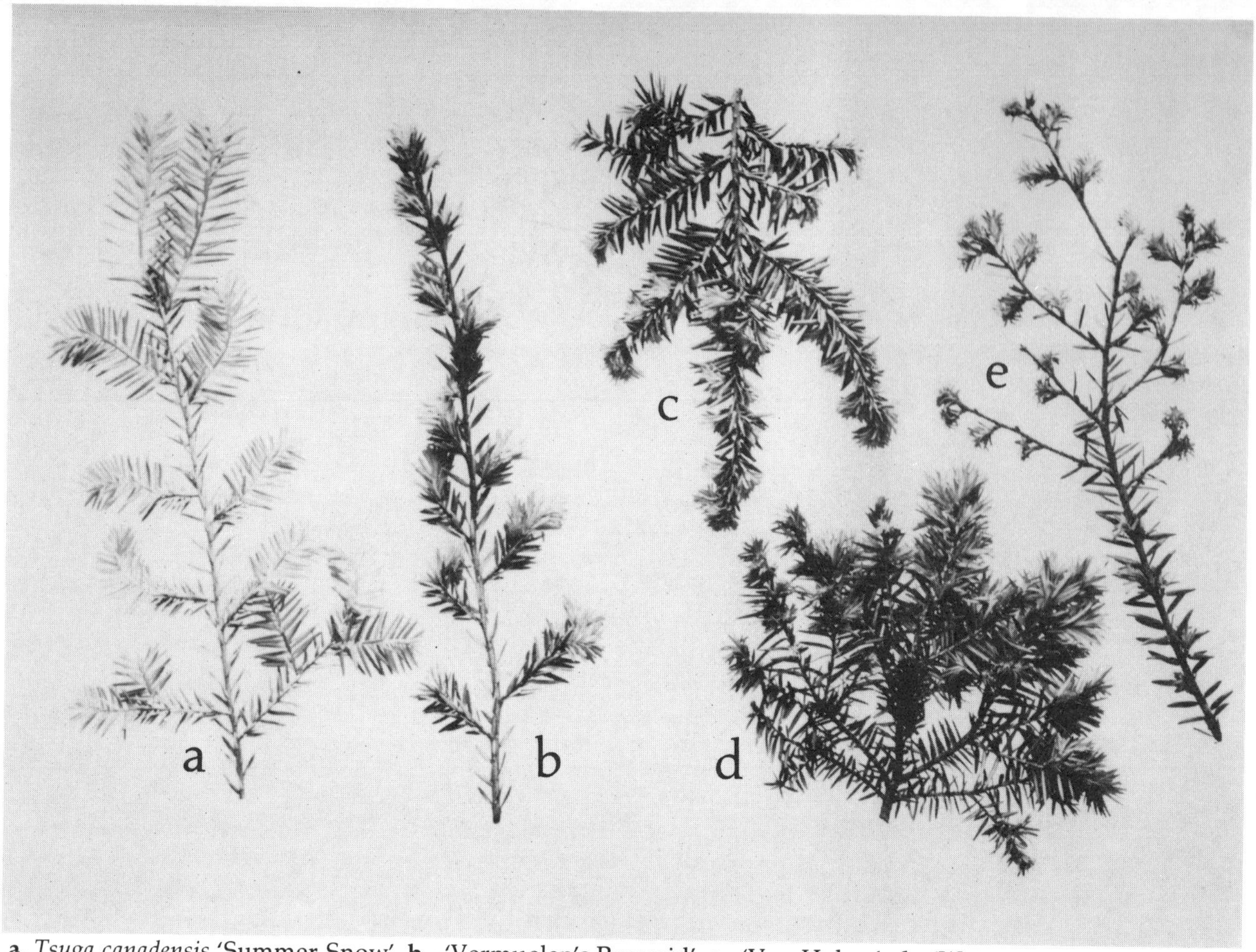

a. *Tsuga canadensis* 'Summer Snow'. **b.**--'Vermuelen's Pyramid'. **c.**--'Von Helms'. **d.**--'Watnong Star'. **e.**--'Youngcone.'

a. *Tsuga caroliniana* 'Ashford'. **b.**--'LaBar Weeping'. **c.**T. *diversifolia*

a. *Tsuga heterophylla* 'Iron Springs'. **b.**--'Morris's Weeping'. **c.**--'Sixes River.'

a. *Tsuga mertensiana* 'Elizabeth'. **b.**--'Mount Arrowsmith'. **c.**--'Mount Hood.'

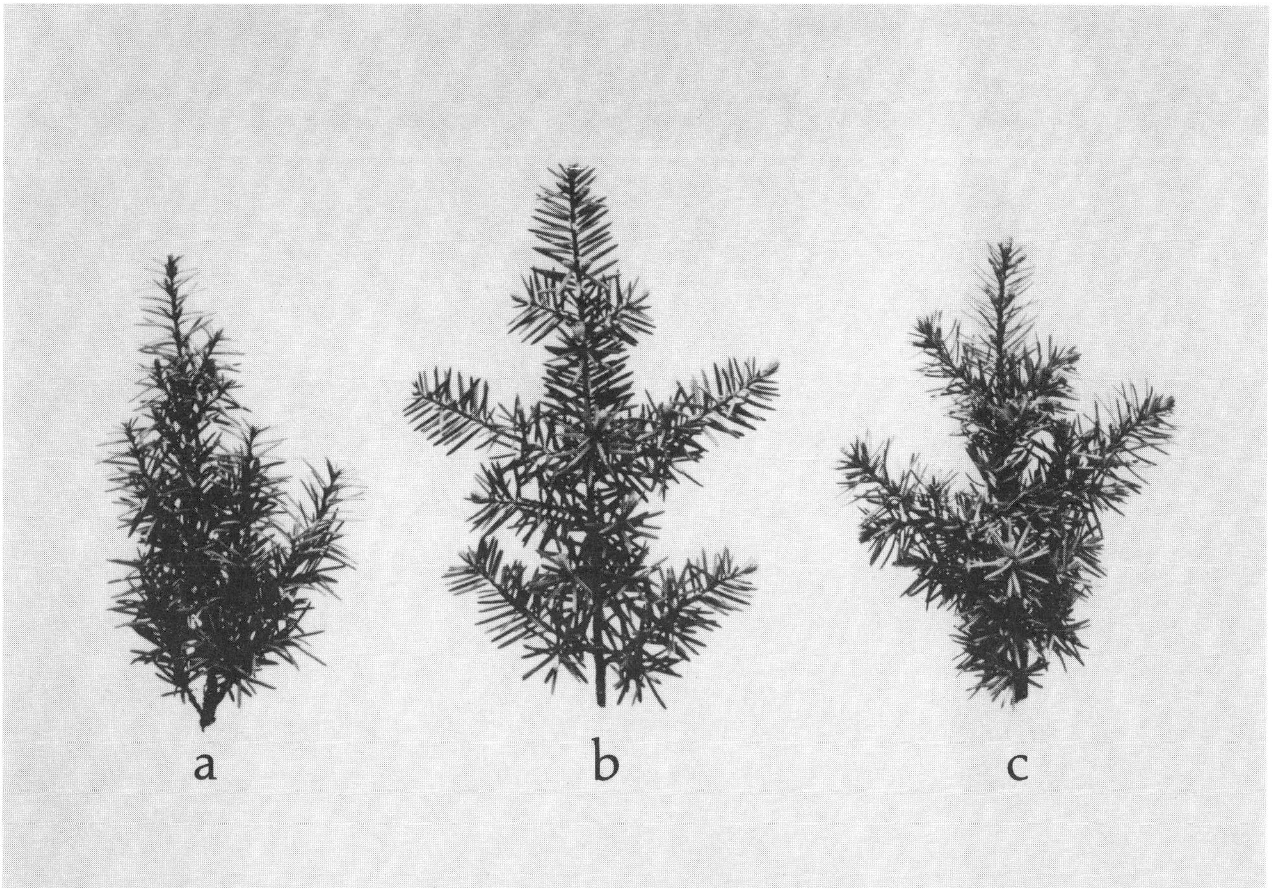

T. caroliniana on a private estate near Boston, Massachusetts.
Photo supplied by Arnold Arboretum, Jamaica Plain, Massachusetts.

6 OTHER SPECIES AND THEIR CULTIVARS *

For convenience of reference, the particulars of *Tsuga canadensis* given in Chapter 4 are repeated here.

THE CANADIAN HEMLOCK; EASTERN HEMLOCK

Tsuga canadensis (L.) Carrière, Traité Conif. 189. 1855.—Sargent, Silva N. Am. **12**: 63, t.603. 1898.
Pinus canadensis Linnaeus, Sp. Pl. ed. 3, **2**: 1421. 1763.
Abies americana Miller, Dict. Gard. ed. 8, *A.* no. 6. 1768.
Pinus-Abies americana Weston, Bot. Univ. **1**: 211. 1770.—Marshall, Arbust. Am. 103. 1785.
Pinus Abies canadensis Muenchhausen, Hausvat. **5**: 223. 1770.
Pinus americana Du Roi, Obs. Bot. 39. 1771.
Pinus mariana Gaertner, Fruct. Sem. **2**: 59, t. 9l, fig. a-g, 1791, exclud. syn. Seligm., Mill.; non Du Roi. 1771.
Pinus pendula Salisbury, Prodr. Stirp. Chap. Allert. 399. 1796, non Aiton. 1789.
Abies canadensis Michaux, Fl. Bor.-Am. **2**: 206. 1803, non Miller. 1768.
Abies pectinata Poiret, Encycl. Méth. Bot. **6**: 523. 1804, non Gilibert. 1792.
Picea canadensis (L.) Link in Linnaea, **15**: 524. 1841.
Tsuga americana (Mill.) Farwell in Bull. Torrey Bot. Club. **41**: 629. 1915.

A tree usually 60-70 feet, and occasionally 100 feet high, with long, slender horizontal lateral branches with pendulous branchlets, forming a broad pyramidal or gradually tapering head. An important, widespread forest tree. Canada from East of the Rocky Mountains to Nova Scotia; U.S.A. Lake States and Appalachian Mountains south to Alabama. It was introduced in 1736 to Britain, where it is common in arboreta and (mostly only the cultivars) in gardens.

BARK. Young trees bright orange-brown with purple-brown scaly ridges. Old trees dark purplish grey with a lattice-work of thick scaly ridges.

CROWN. Young trees broadly conic with obtuse apex and arching slender shoots. Old trees irregularly broad-conic, seldom with a single bole for more than a few feet.

FOLIAGE. Shoot pale brown or cream, with a dense, curly, pale orange pubescence. Second year dark grey or grey-brown.

BUD. Hidden by leaves, broad conic, red-brown.

LEAVES. Pectinate and slightly depressed beneath; above over the shoot and forwards, mid line leaves showing white undersides; broadest at base, tapering slightly to rounded tip. Underside with broad green margins, two white bands and a green midrib which a lens shows to be white centred with a narrow green line each side. Crushed foliage has a sweet lemony scent.

CONE. Ovoid, pendulous, usually numerous, 1.5-2.5 cm, grey-brown; scales rounded, obscurely toothed.

---

*The descriptions throughout this Chapter are based on and the line drawings taken from *Conifers in the British Isles,* A. F. Mitchell (British Forestry Booklet No. 333. Her Majesty's Stationery Office 1972.) by kind permission of Mr. Mitchell and the Forestry Commission. All material thus used is Crown Copyright.

GROWTH. Despite its very slow growth at the seedling stage it grows rapidly, when established, with shoots to 2 feet or more long, but growth in height slows rapidly as the leading shoot tends to lose dominance. There are very few dated trees to show the increase in girth on a single bole but remeasurements show only one approaching one inch per year and the majority less than half this rate. Two dated trees in Britain have, however, exceeded one inch per year.

RECOGNITION. The dark, heavily ridged bark and broad crown at once distinguish this from *T. heterophylla*. The foliage is distinguished from all other *Tsugas* by the line of leaves upside-down along the line of the shoot.

Tsuga caroliniana

Foliage of *Tsuga caroliniana*, Carolina hemlock. The shoot is a shiny pale or reddish-brown and the slender leaves can be quite distant and spread to make a spiky spray. Single leaf (lower right × 1⅓) showing the broad margins to the white stomatal bands on the underside.

TSUGA CAROLINIANA ENGELMANN, THE CAROLINA HEMLOCK

'**T. caroliniana** Engelmann in Bot. Gaz. **6**: 223. 1881.—Sargent, Silva N. Am. 12: 69, t. 604. 1898.
Abies caroliniana (Engelm.) Chapman, Fl. South. U. S. ed. 2, 650. 1883.
Pinus caroliniana (Engelm.) Voss in Mitt. Deutsch. Dendr. Ges. **1907** (16): 94. 1908.

A pyramidal tree 45-60 feet tall with dense, deep-green foliage. It occupies a much more restricted habitat than the Canadian hemlock, occurring only in small, scattered stands, mostly on hillsides, on the mountains of South-west Virginia to Georgia, South East United States. Introduced into Britain in 1886 but still rare, confined to a few collections in South and West England and in Ireland.

BARK. Young trees dark red-brown with large, protruding yellow lenticels. Older trees, purple-grey with wide, shallow, wandering fissures.
CROWN. Open below and inside; densely twigged canopy; irregularly broad-conic or ovoid. The bole has large swellings where the branches arise.
FOLIAGE. Shoot bright red-brown above, pale pink-brown beneath, shining, slightly grooved with pubescence in the grooves. Second year dark red-brown.
BUD. Squat, red-brown, pubescent, 3-4 mm. LEAVES. Sparse and spiky, slender, distant, many standing well above the shoot; 1-1.2 cm long with blunt, rounded tips and two broad bright white bands beneath; midrib with a shiny white centre; very dark green above.
CONE. Long-ovoid, 1.5-2.5 cm, with tall, thin, rounded orange-brown, striated scales, which spread widely when the cone is mature.
RECOGNITION. The slender, distant leaves combined with red-brown shining shoot are distinct.

Although regarded as a more beautiful tree than the Canadian hemlock it is not widely planted, having a reputation for being, in gardeners' language, a bad transplanter. The best specimens I have seen are at the home of the late Professor Sargent at Brookline, Massachusetts. Although the name was not validly published until some years later this species was first discovered by a Dr. Lewis R. Gibbes of Charleston, South Carolina. In the Gray herbarium, Harvard University, Cambridge, Massachusetts, there are two of his letters to Dr. Asa Gray. In the first letter, dated August 2, 1856, Dr. Gibbes claims to have found in 1851 a new species of hemlock occurring in small groups, with different foliage and cones. Apparently Dr. Gray rejected this claim, for in a second letter dated August 19, 1856, Dr. Gray more forcefully insisted on the new species, for which he suggested the name *Pinus* (now *Tsuga*) *laxa*. But although he also reported his find to the Elliott Society of Charleston, South Carolina two years later, he still took no steps to "validly publish" his name and it is now, of course, invalidated by the publication of the present name by Engelmann in 1881.

CULTIVARS OF TSUGA CAROLINIANA

The following cultivars have been selected and named.

- - '**Adams Weeping**' New cultivar. Upright habit with weeping branchlets and more or less asymmetrical growth. The leaves are irregularly disposed. Percy Adams of the Rosenwald Estate in Jenkintown, Pennsylvania, purchased one hundred Carolina hemlock seedlings from a nursery in the South. This one, which was definitely weeping, was presented to Raraflora, Feasterville, Pennsylvania. In 1970 it measured 9 feet high and 7 feet wide.

- - **'Arnold Pyramid'** Wyman, Woody Plant Reg. **1**: 30. 1949, Arnoldia **15**(8): 50. 1955. Wyman described this as a dense pyramidal plant with a rounded top. At the age of thirty-two years it was 25 feet high with a branch spread of 15 feet. It originated as a seedling in the Kelsey Highlands Nursery, East Boxford, Massachusetts in 1938 and was given to the Arnold Arboretum. The writer closely examined this plant in 1973 in the Arnold Arboretum collection and the rate of growth is only half that of the species. The branches are upreaching, showing the lower sides of the leaves. The plant retains a more compact pyramidal shape with leaves more radially disposed. It is a variation distinct from the species but is definitely not a dwarf.

- - **'Ashford'** New name. (*T. caroliniana pyramidalis.* Swartley, Heml. Arb. Bull. 24. 1938). The name *'pyramidalis'* has been rejected because it is a later homonym of a previously described clone of *Tsuga* which is of a different origin. This variant of Carolina hemlock has deviated more in foliage characters than any other thus far observed. It differs from typical Carolina hemlock in having shorter leaves less regularly disposed on the branchlets. There is a tendency for three or four buds to be crowded near the apex of each twig resulting in shorter, denser growth that contrasts with longer, more sparse growth elsewhere; all with a good, dark green colour. The mother plant came to Far Country in 1938 from the Gardens of the Blue Ridge, Ashford, North Carolina. In 1970, at about the age of forty years, it measured 23 feet in height and 13 feet in breadth.

- - **'Compacta'** Hornb. Dwarf Conif. 186. 1923. Hornibrook described this variant as a dwarf, low, and very dense shrub, broader than high; branches horizontal. It had been found among the first cultivated seedlings in the Arnold Arboretum in 1882, but according to Wyman (Arnoldia **15**(8): 50. 1955) there was then no difference between the original plant of 'Compacta' and typical Carolina hemlock. The author agrees with this opinion. It would therefore seem that Hornibrook was misinformed and premature, and that the name should be rejected under Article 42 of the *Cultivated Code* on the

T. caroliniana 'Ashford'. Far Country, Philadelphia, Pennsylvania, in 1945. J.C.S. No. 1793. Photo by F. Ray.

T. caroliniana 'Ashford'. The same cultivar, twenty-three years old, from a cutting. Arnold Arboretum, Jamaica Plain, Massachusetts.

T. caroliniana 'Compacta'. Far Country, Philadelphia, Pennsylvania, in 1945.

grounds that the cultivar "neither does nor did exist."

- - **'Elizabeth Swartley'** New cultivar. This is named from one of two similar plants on the Joseph B. Gable property near Stewartstown, Pennsylvania. These trees are respectively 12 and 15 inch in diameter at breast height and the tree nearest to the house is the clonotype. It has elongated, widely diverging branches. The branches on the lower half of the trunk are declined. Some of the branchlets are wrapped around the branches, partially concealing some of the wood. Other branchlets are free to hang downward, altogether contributing a profile similar to *Cedrus deodara*. It makes a spectacular specimen.

  NOTE: This very elite selection is named in honour of the author's wife, Elizabeth, without whose devotion, patience and profound judgement, especially during the author's illness of later years, this book would not have been possible.

- - **'LaBar Bushy'** New cultivar. A large, loose-growing shrub with multiple stems; lateral branchlets better developed than on typical Carolina hemlock; foliage nearly typical. The original plant came from LaBar's Rhododendron Nursery, Stroudsburg, Pennsylvania in 1938, a gift from Far Country. In 1970 it was 13 feet high and 15 feet wide.

- - **'LaBar Weeping'** Ziegenfuss Nurs. Cat. c. 1967. Name only. This variant of the Carolina hemlock has a weeping habit akin to that of Sargent hemlock. The rate of growth is probably slower. About twelve years ago two similar plants were noticed in a bed of collected seedlings in LaBar's North Carolina nursery. Layne Ziegenfuss of Lehighton, Pennsylvania, for several years propagated this cultivar for LaBar by grafting. The original plants, one of which is the clonotype, are in the custody of Mr. Ziegenfuss.

- - **'Warner Weeping'** New cultivar. The mother plant was given to Far Country in 1945 by Ralph M. Warner, Milford, Connecticut. It is a very graceful plant, about thirty-eight years old in 1970, and measured 33 feet high and 16 inches wide. It is less pendulous than 'Adams Weeping.'

T. caroliniana 'LaFar Bushy'. Far Country, Philadelphia, Pennsylvania. J.C.S. No. 1794. Photo by F. Ray.

T. caroliniana 'LaBar Weeping'. Collection of Llayne Ziegenfuss, Lehighton, Pennsylvania, in 1974.

TSUGA CHINENSIS (FRANCH.) PRITZEL, THE CHINESE HEMLOCK

Tsuga chinesis (Franch.) Pritzel in Bot. Jahrb. **29**: 217. 1900.—Rehder & Wilson in Sargent, Pl. Wilson. **2**: 37 1914.

Abies chinensis Franchet in Jour. de Bot. **13**: 259. 1899.

Abies dumosa var. *chinensis* Franchet, 1. c. 258. 1899, quoad specim. Farges.

Tsuga Brunoniana var. *chinensis* (Franch.) Masters in Jour. Linn. Soc. Lond. Bot. **26**: 556. 1902, p. p.—Beissner in Mitt. Deutsch. Dendr. Ges. **1903**(16): 67. 1903; Handb. Nadelh. ed. 2, 84. 1909, p. p.

Tsuga yunnanensis (Franch.) Masters in Gard. Chron. ser. 3, **39**: 236, fig. 93. 1906, p. p.

Tsuga formosana Hayata in Gard. Chron. ser. 3, **43**: 194. 1908; in Repert. Sp. Nov. Reg. Veg. **8**: 366. 1910.—Flous in Trav. Lab. For. Toulouse, II, 4,3: 84, fig. (Rév. Tsuga) 1936; in Bull. Soc. Hist. Nat. Toulouse, 71: 398, fig. 1937.

A tree to 120 feet tall, so far not widely in cultivation. Introduced in 1900 by E. H. Wilson to the Veitch nursery at Exeter, England, and to the Arnold Arboretum in 1909, but still very rare both in this country and in Britain, Central and West China.

BARK. Dark orange-brown coarsely flaked with patterns of green-grey flakes.

CROWN. Usually seen as a strong bush with ascending branches and vertical axis or a small conical tree, of a distinct yellowish-green with protruding shoots.

FOLIAGE. Shoot buff above, shiny cream-white beneath with rich brown pulvini; minutely pubescent, nodding.

BUD. Ovoid, dark red-brown, sometimes with green leaves protruding.

LEAVES. Arranged rather as in *T. heterophylla* but distinctively evenly spaced and long leaves separated by several short leaves, 0.5-1.5 cm, light fresh green above, parallel sides to a blunt, slightly notched tip; two narrow greenish bands beneath.

CONE. Ovoid stalked, 2.5 cm, shining yellowish-brown, incurved rounded scales.

RECOGNITION. The crown differs at once from *T. heterophylla* but the foliage is similar, differing in the apparently glabrous shoot, pale upper surface of the leaf, and, most distinct, the greenish bands beneath. *T. forrestii* (q.v.) is perhaps only a variety.

T. chinensis at Swarthmore College, Swarthmore, Pennsylvania, in 1972.

Bailey (*The Cultivated Evergreens* 266. 1923) says of it "A handsome species promising well in suitable localities: hardy in sheltered positions as far north as Massachusetts." From a distance it resembles *T. sieboldii* but it is more decorative and has longer leaves.

The following botanical varieties have been recognised, but no cultivars seem to have been recorded.

 T. chinensis var. *daibuensis* S.S. Ying, Bull. Exper. Forest. Nat. Taiwan Univer. **114;** 150. 1974, Quart. Journ. Chinese For. 8(3): 98. 1975—Taiwan.
 T. chinensis var. *formosana* (Hay) Li et Keng ex S.S. Ying, Quart. Journ. Chinese For. 8(3): 97. 1975. Formosa.
 T. chinensis var. *oblongisquamata* C. Y. Cheng et al. Acta Phyto. Sinica 13(4): 83. 1975. People's Republic of China.
 T. chinensis var. *robusta* C. Y. Cheng et al. Acta Phyto. Sinica 13(4): 83. 1975. People's Republic of China.

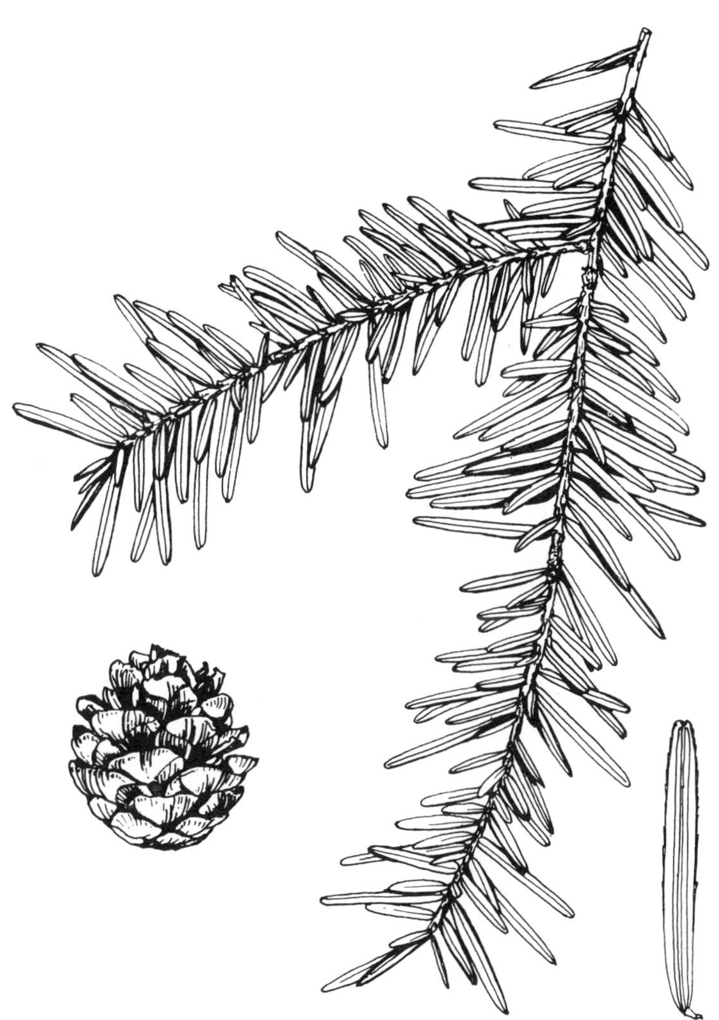

Tsuga chinensis

Foliage and ripe cone of *Tsuga chinensis,* a Chinese hemlock which differs from the Western hemlock mainly in the pale green upper side of the leaf which has only greenish-white bands beneath. (detail × 1⅓).

TSUGA DIVERSIFOLIA (MAXIMOWICZ) MASTERS, NORTHERN JAPANESE HEMLOCK

Tsuga diversifolia (Maxim.) Masters in Jour. Linn. Soc. Lond. Bot. **18**: 514. 1881.
Abies Tsuga B. *nana* Endlicher, Syn. Conif. 83. 1847.
Tsuga Sieboldii B. *nana* Carrière, Traité Conif. 186. 1855.
Tsuga Tsuja var. *nana* (Carr.) A. Murray in Proc. Hort. Soc. Lond. **2**: 509. 1862.
Abies diversifolia Maximowicz in Bull. Acad. Sci. St. Pétersb. **12**: 229 (in Mél. Biol. **6**: 379) 1868.
Tsuga Araragi var. *nana* (Endl.) Sargent, Silva N. Am. **12**: 60, in nota. 1899.
Pinus araragi var. *diversifolia* Voss in Mitt. Deutsch. Dendr. Ges. **1907**(16): 93. 1908.
Pinus Araragi var. *nana* Voss in Putlitz & Meyer, Landlex. **4**: 773. 1913.

A pyramidal tree to 25 m, making a more attractive specimen than its compatriot, *T. sieboldii*. Japan, in Central and Southern Honshiu. Introduced in 1861. Rare; confined to collections and a few large gardens.

BARK. Dark orange-brown with pink fissures and with ridges shallowly cracked or flaking vertically.
CROWN. Almost invariably a multitude of straight stems from the ground, forming a broad and very dense, low-domed crown.
FOLIAGE. Shoot orange beneath with fine erect pubesence visible with a lens; second year dull brown. Leaves pectinate below, tips slightly curved down, very regular and closely set on side-shoots, oblong with broad rounded and notched tip, 1 × 0.2 cm, deep glossy green above; two broad bright white bands beneath with green midrib and margins clearly defined.
BUDS. Pear-shaped, glabrous, dark-brown, the smaller buds globose; apex very blunt. About 3 mm.
FLOWERS AND CONE. Male flowers among the leaves near the tips of side-shoots, bright red-brown as buds. Cone ovoid, 2 cm, shiny brown.
GROWTH. Growth in height is very slow, dated trees averaging little more than 6 inches a year at best. Growth in girth is difficult to assess as single stems are so rare but it is in any case very slow.
RECOGNITION. The short, broad, blunt leaves are similar only to *T. sieboldii,* from which they differ in being broader, more regularly placed, deeper, glossier green above, and whiter beneath. The shoot is also much brighter and darker and minutely pubescent. The crown is broader and more dense and bushy.

Charles Jenkins (*Hemlock Arboretum Bulletin* 30. 1940) records one year when *Tsuga diversifolia* came out with a glorious display of pale lavender flowers. The tree was covered with bloom which lasted for a week or ten days. This must put this species into a unique position within the genus. Another of its striking characteristics, the white stomatic bands, are so conspicuous that its popular name in Japan is Rice Tree. It is traditionally the first of the hemlocks to come into leaf in the Spring, *T. sieboldii* being the last. This should be very convenient for amateur botanists if mixed stands of the species occur anywhere in Japan!

CULTIVARS OF TSUGA DIVERSIFOLIA

The following cultivars have been selected and named.

- - **'Gotelli'** New name (*T. sieboldii nana* Hort. Gotelli, NON Carr.). The mother plant in the Gotelli collection in the U.S. National Arboretum, Washington, D.C. (Accession No.

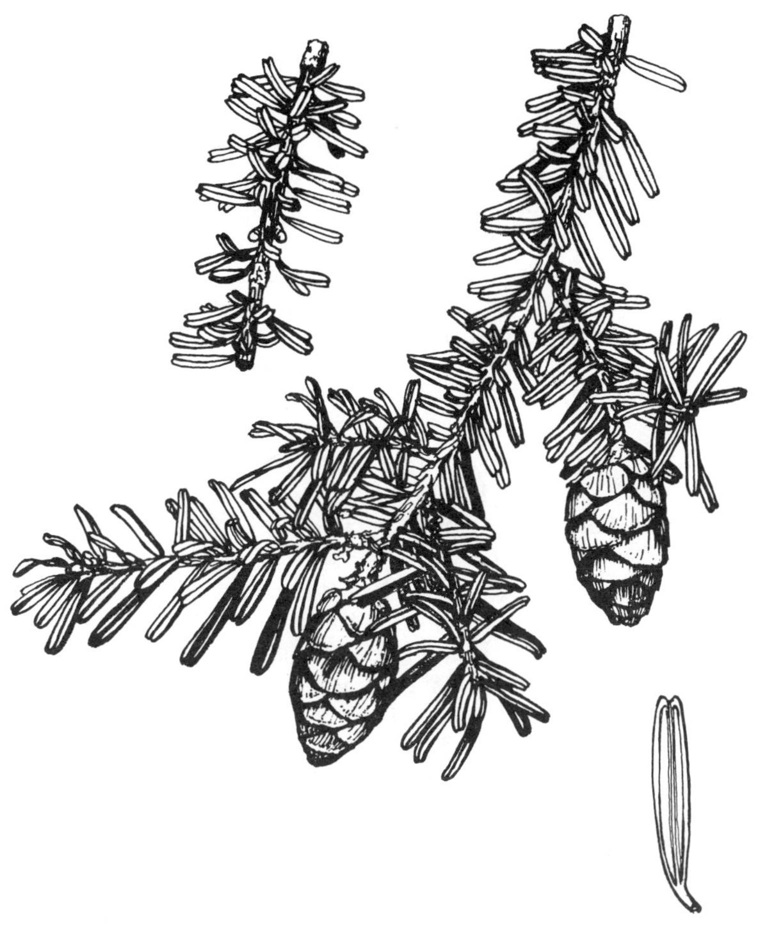

Tsuga diversifolia

Foliage and cones of *Tsuga diversifolia*, a hemlock from Southern Japan. The shoot is bright orange, very finely pubescent, and from beneath (upper left) contrasts strikingly with the brilliant white of the underside of the leaves (lower right × 1⅓). The foliage is much brighter, more regular and uniform than that of *T. sieboldii*.

20862) is dead, but had been propagated. It was 2 feet high and 3 feet wide in 1970. According to Mr. Gotelli it was about thirty years old and 2 feet high in 1962 when received from the Thompson Nurseries in New Zealand. The habit is dense and irregular with annual growth less than 1 inch; winter buds moderately crowded; leaves shorter than typical *T. diversifolia. Tsuga sieboldii* var. *nana* being a synonym of *T. diversifolia* (Maxim.) Masters, a cultivar name has had to be found. This name is chosen in honour of William T. Gotelli, of South Orange, New Jersey, who in 1962 donated a large collection of dwarf and other conifers to the U.S. National Arboretum, Washington, D.C.

- - **'Manifold'** New name (*T. sieboldii fastigiate* Hort. Gotelli). Named for another plant in the Gotelli collection at the U.S. National Arboretum, Washington, D.C. (Accession No. 21663), a flat-topped tree 6 feet high and 8 feet wide. The horizontal main branching is quite irregular and open, with close compact growth on the branches arranged in whorls. It is not fastigiate. The foliage is so dense as to give the effect of weighing down the branches. This was another acquisition from the Thompson Nursery in New Zealand, one of three selections from a large group of seedlings raised on that nursery. The third plant was labelled *Tsuga sieboldii compacta* but it cannot be traced and was never propagated, so the name may be discarded.

A note from Mr. Gotelli at the National Arboretum states that this tree was fifty to sixty years old in 1962 when it was 5 feet high. It would seem more logical if the ages were reversed; 'Gotelli' fifty to sixty years old and 'Manifold' about thirty years old in 1962.

- - **'Medford Lake'** New cultivar. A tree of irregular, semi-compact habit with no definite leader. Slow growing with lateral growth 1-2 inches, leaves 9-11 mm long and more irregularly disposed than in the species. The origin is not recorded. The oldest specimen known is in the planting of a house belonging to the Medford Nursery, Medford, New Jersey. It has been propagated and distributed by Edward Harbaugh, Taunton Lake, Medford, New Jersey. A plant propagated from a cutting about fifteen years ago was about 4 feet 6 inches high in 1974.

- - **'Thompson'** New cultivar. A plant of *T. diversifolia* (Accession No. 1304-60) was observed in the Arnold Arboretum in 1973. This measured 4 feet high and 10 inches wide. Propagating material had come from the Thompson Nurseries, Kinderhook, New York. It had been mislabeled *T. sieboldii.* Two other variants of presumably the same species had been sent at the same time, but these had probably failed.

- - 'Totten' Provisional name. Walter Kolaga, proprietor of the former Mayfair Nurseries, Nichols, New York, found this variant. In a letter he described it as a round-top dwarf about 3 feet high and about forty years old, in the garden of Edgar Totten at Ho-ho-kus, New Jersey. I do not know whether or not it is in cultivation.

T. diversifolia at Far Country, Philadelphia, Pennsylvania, in 1938.

T. diversifolia, 'Gotelli'. U.S. National Arboretum, Washington, D.C., in 1970. J.C.S. No. 1789.

T. diversifolia. Foliage detail with coves.

T. diversifolia 'Manifold'. U.S. National Arboretum, Washington, D.C. Accession No. 21663.

TSUGA DUMOSA (D. DON) EICHLER, HIMALAYAN HEMLOCK.

Tsuga dumosa (D. Don) Eichler in Nat. Pflanzenfam. II. **1**: 80. 1809.
Pinus dumosa D. Don, Prodr. Fl. Nepal. 55. 1825.
Pinus Brunoniana Wallich, Pl. As. Rar. **3**: 24, t. 247. 1832.
Abies Brunoniana (Wall.) Lindley in Penny Cycl. **1**: 30. 1883.
Abies dumosa (D. Don) Loudon, Arb. Brit. **4**: 2325, fig. 2233, 2234. 1838.
Picea Brunoniana (Wall.) Wenderoth, Pflanz. Bot. Gärt. I. Conif. 60 .1851.
Tsuga Brunoniana (Wall.) Carrière, Traité Conif. 188. 1855.
Abies cedroides Griffith mss. ex Carrière, Traité Conif. 188. 1855.
Micropeuce Brunoniana Spach ex Carrière, Traité Conif. ed. 2, 247. 1867, pro syn.
Tsuga Brunoniana var. *typica* Patschke in Bot. Jahrb. **48**: 767. 1913.
Tsuga leptophylla Handel-Mazzetti in Anzeig. Math.-Nat. Kl. Akad. Wiss. Wien, **1924**: 83
  (Pl. Nov. Sin. Forts. **25**: 3. 1924.

A pyramidal tree to 40 m high, but so far not widely in cultivation. Himalaya from Kumaon to Bhutan. Introduced in 1838. Rare both in this country and in Britain.

BARK. Pink-brown and very scaly, becoming in old trees very like that of an old larch; heavily ridged, shallowly fissured, very scaly and pink.
CROWN. Where the tree grows well, a tall, irregular tree with widely spreading branches and several conic tops; the shoots pendulous. Elsewhere an ovoid bush or straggling thin tree.
FOLIAGE. Shoot pale pinkish-brown with a scattered fine pubescence.
BUD. Globose, pubescent.
LEAVES. Distant, hard and rigid, pointing forward along the shoot, all on the upper side; 1-3 cm, long, tapering from near the base or about half way to a rounded end, grooved above; two very broad and very white bands beneath, with scarcely any green margin.
CONE. Ovoid, 2-3.5 cm, scales rounded, striated and shining (Dallimore and Jackson).
RECOGNITION. The big, forward-pointing leaves on pendulous shoots are distinctive.

There are no recorded cultivars of *Tsuga dumosa.*

T. dumosa at "Graywood Hill", Surrey, U.K.

Tsuga dumosa

Tsuga dumosa, Himalayan hemlock, has the largest leaves of all the hemlocks. They are hard, with broad white bands beneath (detail ✕ 1⅓) and the foliage is pendulous.

TSUGA HETEROPHYLLA (RAF.) SARGENT.
WESTERN HEMLOCK.

Tsuga heterophylla (Raf.) Sargent, Silva N. Am. **12**: 73, t. 605. 1898.
Pinus canadensis sensu Bongard in Mém. Acad. Sci. St. Pétersb. **2**: 163 (Obs. Végét. Sitcha, 54. 1832, non Linnaeus 1767.
Abies heterophylla Rafinesque, Med. Fl. **2**: 182. 1830, nom.; in Atlant. Jour. **1**: 119. 1832.
Abies microphylla Rafinesque, l. c. 1830, nom.; l. c. 1832.
Abies Mertensiana sensu Gordon, Pinet. 18 (1858), non Lindley & Gordon. 1850.
Abies taxifolia Jeffrey ex Gordon, l. c. (1858), pro syn.; non Poiret (1804), nec Desfontaines. 1804.
Picea Mertensiana hort. gall. ex Gordon, Pinet. 18. 1858, pro syn.
Abies Albertiana A. Murray in Proc. Hort. Soc. Lond. **3**: 149, fig. 6, 7, 9, 11, 13, 15. 1863.
Abies Bridgesii Kellogg in Proc. Calif. Acad. Sci. **2**: 37. 1863.
Tsuga Mertensiana Carrière, Traité Conif. ed. 2, 250 1867, quoad descript., non *Pinus Mertensiana* Bongard. 1832.
Tsuga Albertiana Sénéclauze, Conif. 18. 1867.
Pinus Mertensiana sensu Parlatore in De Candolle, Prodr. **16, 2**: 428. 1868, non Bongard. 1832.
Pinus Pattoniana W.R. McNab in Proc. Roy. Irish Acad., ser. 2, **2**: 211, 212, t. 23, fig. 2. 1875, non Parlatore. 1868.
Abies Patonii Jeffrey msc. ex W. R. McNab in Jour. Linn. Soc. Lond. Bot. **19**: 208. 1882, non Gordon. 1858.

A narrowly pyramidal tree to 60 m, with usually pendulous branches, usually broader when young. North-west America from the coast of South West Alaska to Northern California; inland in Southern-central British Columbia. It was introduced into Britain in 1851, where it does well and is the only species to have been widely planted. Frequent in most areas in large gardens and some smaller. Very common in policies in Scotland; frequent as plantation tree especially under or replacing old hardwoods.

BARK. Young trees reddish purple-brown cracking into circular flakes. The flakes later come away and leave red-purple patches. Old trees: dark purple-brown, sometimes grey; fissured finely into irregular, broken ridges.
CROWN. Always distinctively narrow and conic to a slender, gracefully arched and drooping leading shoot. Branches ascending, arching gently out to the tips, with narrow plumes of dense foliage on their undersides partly pendulous; dark and dense.
FOLIAGE. Shoot reddish brown above, creamy-white beneath, ribbed; dense curly, long pale brown pubescence; second year dull brownish grey.
BUD. Hidden by leaves; narrowly conic, pale brown, pubescent. Side-buds in May prominent, globular and white before expanding.
LEAVES. In two sets, short nearly upright, 0.5 cm, above the shoot; long 1.5-1.8 cm × 0.2 cm, flat pectinate each side of the shoot; parallel-sided to a blunt apex, pale green when fresh, soon blackish-green; two broad blue-white bands beneath almost obscuring midrib; very narrow green margins.
FLOWERS AND CONE. Male flowers clustered among the leaves on the ends of short shoots, crimson before shedding pollen in late April and early May; the pollen abundant. Cone 2-2.5 cm long, ovoid, pale green with few scales, ripening dark brown.
GROWTH. Growth in height is for many years remarkably vigorous on a wide variety of soils from deep neutral loams to dry acid sands. Leading shoots of 4 feet 6 inches are quite frequent. Growth starts downwards from the hanging tip in mid-May; attains 3-5 inches a week in July when most of the growth of the previous year straightens up but leaves the new growth still hanging or at a strongly downward angle. Growth continues until early or mid-September, by which time the basal portion of the new growth is upright. Trees have been found which have grown 60 feet in 20 years, and in

Tsuga heterophylla

Foliage and cone of *Tsuga heterophylla*, Western hemlock, with (below) seed ($\times$ 1⅓) and underside of leaf ($\times$ 1⅓) showing broad white stomatal bands.

Scotland, growth on trees well over 100 feet tall may still be above 1 foot a year. Growth in girth is sometimes extremely rapid.

GENERAL. This species establishes most rapidly under light shade. It is able, although it is inadvisable to let it, to grow through over-hanging branches, since the fragile new leading shoot is hanging below and the contact with the branches is made by woody 1-2 year old wood, as that part of the leader straightens. Although crops are not particularly stable, individual trees are extraordinarily wind firm. It is not a species to grow in great exposure nor on deep peats, but it will grow in both conditions combined, making bushy, multi-stemmed growth for several years and then a single leading-shoot emerging, and growing steadily if slowly. The tree is susceptible to butt-rot (*Fomes annosus*) and is liable to much loss from this fungus when it is planted in old hardwood areas. There has thus been a recent decrease in such planting.

RECOGNITION. The single-boled, long-spired crown and drooping leading shoot are distinct from all other hemlocks and all other trees, although the very different deodar (*Cedrus deodara*) shares these particular features. The foliage differs in layout from other hemlocks except *T. chinensis,* but unlike that species, the leaves are broadly banded bluish-white beneath.

Many fine specimens in Britain are recorded by Mitchell (*Conifers in the British Isles* 310. 1972).

CULTIVARS OF TSUGA HETEROPHYLLA

The following cultivars have been selected and named.

- - **'Argenteo-variegata'** (Beissn.) Silva-Tarouca, Nadel. 294. 1923; *T. mertensiana* f. *argenteo-variegata* Hort. ex Beissn. Nadel. ed 2. 94. 1909. Branchlets appear as though powdered white when young. Probably no longer in cultivation.
- - **'Conica'** den Ouden, Coniferen 362. 1949. A conical to ovoid shrub, dense, to 10 feet high; branches erect, tips drooping. A very conspicuous form, raised about 1930 from seed in the arboretum of von Gimborn, Doorn, Netherlands. In the early 1950s P. den Ouden sent the author a photograph of the same plant under the name 'Globosa.' This variant is unknown in the United States.
- - **'Dumosa'** den Ouden, Coniferen 362. 1949. A dwarf, bushy, slow-growing shrub; branches spreading, stout, but short; sprays numerous, very short; leaves 12-15 mm long and ± 2 mm broad, green above, and with two stomatic bands beneath. Selected in 1934 in the nursery of the firm Bijvoet, Oosterbeck, near Arnheim, Netherlands. The original plant was about 2 feet high and wide in 1944. This cultivar is unknown in the United States.
- - 'Flaccida' Mr. H.S. Newins wrote (Journ. For. **20:** 272. 1922) of a remarkable weeping hemlock for which he suggested this name and commented on the difficulty of propagating it, but in February 1940 Mr. Jenkins received a letter from Thornton Munger, Portland, Oregon, reporting that the original weeping tree was destroyed by the logging of Douglas fir. There is no record of its being propagated. Article 10 of the Cultivated Code defines a cultivar as "an assemblage of plants", so a tree that has never been propagated cannot be regarded as a cultivar, and Article 42 states that a cultivar name must be rejected if the cultivar of which it purports to be the name "neither does nor did exist."

There would therefore appear to be no reason to retain this name.
- - **'Greenmantle'** Hillier, Manual 531. 1971. A graceful, tall, narrow tree with pendulous branches which originated at Windsor Great Park, Berkshire, England. The foliage is typical; leaves are parallel-sided 15 mm × 1.5 mm, rounded or blunt-tipped. The shoots are very thin, dark grey-brown.
- - **'Iron Springs'** Registered name. Witt, Assoc. Amer. Bot. Gar. Arb. Bull. 9(2): 47. 1976. A slow-growing shrub with upright irregular habit; the original plant found in 1964 is now about 6 m tall and 5 m wide. The older wood is greyish-brown; the new growth which may vary from 1 to 5 cm. in length is pale tan in colour. The leaves are much shorter than the type western hemlock, ranging from 2-3 mm to nearly 1 cm. They are dark green above and silvery-glaucous below.

This plant was discovered growing near Iron Springs, a resort on the Pacific Coast of Washington State, by Mrs. Ainsworth Blogg of Seattle. Mrs. Blogg gave the original plant to the University of Washington Arboretum in 1969.
- - 'Krause's Compacta' Provisional name. A seedling that appeared in the Northwest Ground Covers Nursery, Woodville, Washington, U.S.A. Now, seven or eight years later it is a prostrate mat 2 feet in diameter but only 3 inches high, with a habit similar to *Tsuga canadensis* 'Cole'. The leaves are unusually small. It was first propagated by John Krause of Canyon View Nursery, Entiat, Washington, U.S.A. In hot, dry climates it may need some shade. Unfortunately the suggested name is illegitimate, so a more acceptable name should be registered.

- - **'Laursen's Column'** Hillier, Manual 531. 1971. A striking tree of loosely columnar habit recalling *Podocarpus andinus* in general appearance. A seedling found by Mrs. Asger Laursen in 1968. Its allegiance to this species is open to doubt. The mother plants in the Hillier Arboretum, Romsey, Hants. consist of a group of clonally propagated trees 4 m high by less than 50 cm across. The leader is erect but the branches are strongly pendulous. The foliage is dense and regular, the inconspicuous stomata beneath the leaves not affecting the overall color, which is very dark green. Its allegiance to this species is perhaps open to doubt, but it should develop into a spectacular tree.
- - **'Morris's Weeping'** Provisional name. This was found as a seedling by Rollin D. Morris in the Willapa Bay National Wildlife Park, Washington, U.S.A. The mother tree is now 10 m high, with strongly pendulous branches. The leaves are 8-17 mm long, imperfectly two-ranked. It bears cones 2-2.5 cm long, ripening to red clay brown. This cultivar has been introduced to the trade by several nurseries in Oregon. A more acceptable name should be registered.
- - **'Sixes River'** New cultivar. This was named for the location in Oregon where it was found by Ronald Gourley of Blue Diamond Nursery, Winston, Oregon, U.S.A. One limb only was cascading down from a normal tree 75 feet high whose branches developed a very unusual and quite pronounced curve at each joint. Three-year-old plants raised from cuttings are 3 feet tall and wide.

T. heterophylla 'Epstein Dwarf'. Collection of Joel Spingarn, Baldwin, Long Island, New York, in 1973.

T. heterophylla 'Iron Springs'. Photo supplied by the Washington Arboretum, University of Washington, Seattle (Joseph Witt, director).

TSUGA × JEFFREYI (HENRY) HENRY. JEFFREY HEMLOCK.

Tsuga × **Jeffreyi** (*T. heterophylla* × *mertensiana*) (Henry) Henry in Henry & Flood in Proc. Roy. Irish Acad. 35B: 55. 1919.
Tsuga Pattoniana var. *Jeffreyi* Henry in Elwes & Henry, Trees Gr. Brit. Irel. 2: 231. 1907.
Tsuga mertensiana var. *Jeffreyi* Schneider in Silva Tarouca, Uns. Freil.-Nadelh. 294. 1913.

Formerly regarded as a variety of *T. mertensiana*, now regarded as a hybrid between *T. mertensiana* and *T. heterophylla*. If this be so, the second parent shows remarkably little in the hybrid; the bark, crown and cone being almost exactly those of *T. mertensiana*. A tree with a curious and doubtful history in cultivation. It was first noticed in Edinburgh Royal Botanic Garden from seed sown in 1851 deriving from near Mount Baker in British Columbia. Another tree was raised there more recently among Rhododendron seeds from the Selkirk Mountains and again one was grown in Ireland from seedlings dug up in Vancouver Island. It was thus unknown in the wild yet occurred three times in the British Isles. In 1967 or 1968 however, Mr. J. Duffield found a hybrid swarm of this parentage growing on the east side of White Pass, south of Mount Rainier, Washington. The origin of the few British specimens is unknown but presumably they were raised from the first tree at Edinburgh. They do not appear to be grafts. There is some variation in these, towards *T. mertensiana*.

BARK. Deep orange-brown or blackish, finely fissured and very scaly.
CROWN. Young plants ovoid, many ascending shoots competing with the leading shoot. Leading shoot bent to one side at the tip. Trees narrowly conic, rather open; descending branches. Layered branches at Lydhurst.
FOLIAGE. Shoot pale brown with a sparse, long pubescence. Leaves nearly pectinate, sparsely set perpendicular to shoot, some pointing backwards, 1-1.5 cm, very slender, and flat, grooved above, light greyish yellow-green both sides.
CONE. Cylindric-oblong to 5 cm (Dallimore and Jackson). It is rare in cultivation in the United States.
RECOGNITION. A tree like *T. mertensiana* but instead of thick, blue-grey leaves all round the shoot and lying forwards, slender, flat yellowish-grey-green leaves, pectinate, perpendicular and some pointing backwards.

There are no recorded cultivars of *Tsuga* × *jeffreyi*.

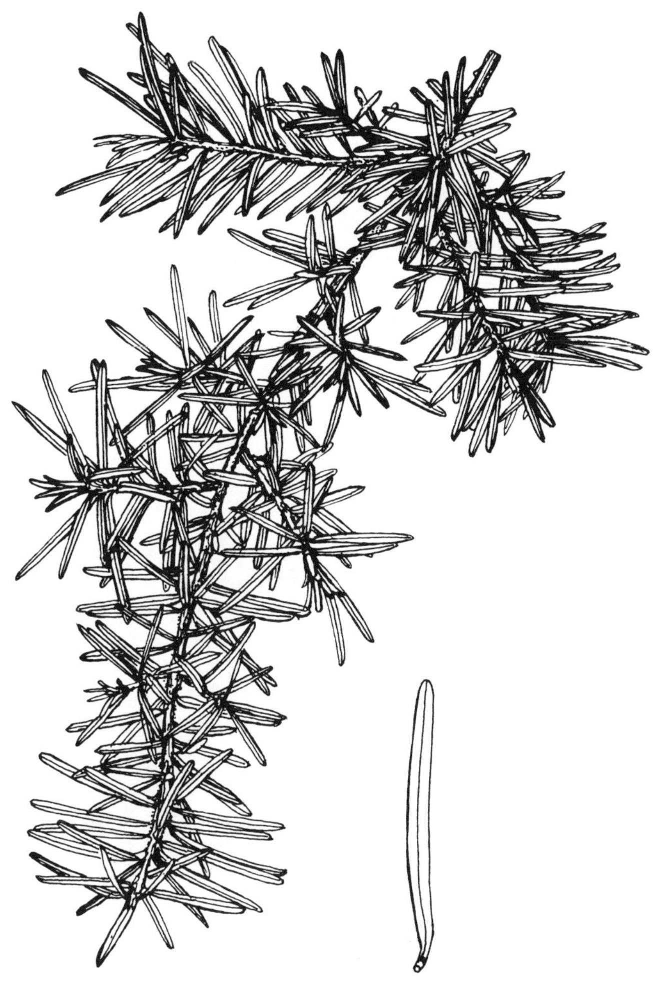

Tsuga × jeffreyi

The hybrid *Tsuga × jeffreyi*, unknown in the wild until recently, has grey-green leaves, more slender and spreading than those of the one of its parents it most resembles, *T. mertensiana*. They are the same colour both sides and grooved along the upper side (right × 1⅓).

TSUGA MERTENSIANA (BONG.) CARRIÈRE.
MOUNTAIN HEMLOCK.

Tsuga mertensiana (Bong.) Carrière, Traité Conif. ed. 2, 250. 1867, quoad basonym., descript. exclud.—Sargent, Silva N. Am. **12**: 77, t. 606. 1898.

Pinus Mertensiana Bongard in Mém. Acad. Sci. St. Pétersb. **2**: 163. 1833. (Obs. Végét. Sitcha 54. 1832) 1833.

Abies Mertensiana (Bong.) Lindley & Gordon in Jour. Hort. Soc. Lond. **5**: 211. 1850.

Abies Pattoniana A. Murray, Rep. Oreg. Exp. 1, t. 4, fig. 2. 1853.

Picea Californica Carrière, Traité Conif. 261. 1855.

Abies Hookeriana A. Murray in Edinb. New Phil. Jour. n. ser., **1**: 289, t. 9, fig. 11-17. 1855.

Abies Williamsonii Newberry in U.S. (War Dept.) Rep. Expl. Surv. Railr. Mississippi Riv. Pacif. Oc. **6,3**: 53, fig. 19, t. 7. 1857.

Abies Pattonii Jeffrey ex Gordon, Pinet. 10. 1858; Suppl. 6. 1862.

Tsuga Williamsoni Vos, Bered. Woordenb. Heest. Conif. 181. 1867.

Tsuga Hookeriana Carrière, Traité Conif. ed. 2, 252. 1867.—Flous in Bull. Soc. Hist. Nat. Toulouse, **71**: 414, fig. (Rév. Tsuga, 100. 1936). 1937.

Pinus Pattoniana (Murr.) Parlatore in De Candolle, Prodr. **16,2**: 429. 1868.

Tsuga Roezlii Carrière in Rev. Hort. **1870**: 217. 1870.

Picea (Tsuga) Hookeriana (Carr.) Bertrand in Ann. Sci. Nat. Bot. sér. 5,**20**: 89. 1874.

Pinus Hookeriana W. R. McNab in Proc. Roy. Irish Acad. ser. 2,**2**: 211, 212, t. 23, fig. 1. 1875.

Tsuga Pattoniana (Jeffrey) Engelmann in Gard. Chron. n. ser. **12**: 756. 1879.

Hesperopeuce Pattoniana Lemmon in Bienn. Rep. Calif. State Board For. **3**: 126, t. 9. 1890.

Tsuga Pattoniana var. *Hookeriana* (Murr.) Lemmon, W. Am. Cone-bear. 54. 1895.

Hesperopeuce Mertensiana (Bong.) Rydberg in Bull. Torrey Bot. Club, **39**: 100. 1912, exclud. syn.

A pyramidal tree 30-50 m high, shrubby only at high altitudes. Islands and coast of South Alaska and British Columbia; Washington, few in Olympic Mountains; abundant high on the Cascades through Oregon to the Sierra Nevada, California. Introduced in 1854. Uncommon in Britain, but in many large gardens, particularly in northern parts, and in some small gardens in areas like South West Surrey, where many conifers are grown.

BARK. Brownish-orange, fissured finely vertically and flaking into small rectangular scales.

CROWN. Narrowly conic and pointed, usually a bushy base where growth was slow, tending to be narrowly columnar with a long spire where growth has accelerated; short level or depressed branches and variably pendulous pale blue-grey foliage; very dense.

FOLIAGE. Shoot, pale shining brown with a short pubescence; pulvini rich red-brown abruptly changing to greenish-white. Leaves arising all round the shoots and with many short shoots vertical above the spray, the appearance is there like the whorled leaves of a cedar. Leaves lying well forwards, thick, 1.5-2 cm long, parallel sided to a rounded tip, uniformly glaucous but variably blue, all round.

CONE. (ex-California) 7 × 3.5 cm (open) cylindric, tapered slightly, fawn-pink stained purple, striated. In these countries soon dark brown; clustered around the apex.

GROWTH. At first growth is slow with leading shoots of at most one foot in length and this persists for many years. Later, especially in Scotland, growth is quite rapid and the largest trees are growing rapidly in height and girth. Although a mountain species, the best trees in Britain are all on rich soils in lowland arboreta, mainly in damp areas with cool summers, and high altitude plantings whilst remaining healthy, have grown but slowly.

RECOGNITION. The narrowly columnar-conic crown of pendulous grey foliage is quite distinct. The thick, grey-blue leaves all round the shoots and pointing forwards are also distinct among all hemlocks (and other trees). The colour of the foliage varies in the whitish blue shade and the brightest are grown as cv. Argentea whilst many almost

Tsuga mertensiana

Foliage and cone of *Tsuga mertensiana,* Mountain hemlock. This aberrant species has a cone like a spruce and leaves radiating all round the shoot. The short shoots can look like spurs, and since the leaves are grey or blue-grey, there is a passing resemblance to the foliage of *Cedrus atlantica* 'Glauca'. Inset leaf $\times$ 2.

normally coloured are grown as cv. Glauca but the intermediate colours are too many to establish these as separate cultivars, and occur in adjacent trees in Oregon.

This species is widely acclaimed as the most beautiful of all the hemlocks and not without good reason. Selections with glaucous foliage are always attractive. But curiously, it is not as popular as a landscape plant in the western United States; perhaps this is because of its slow growth in the nursery, making it an uneconomic tree to grow.

In this species only one variation has been widely recognised by botanists and the number of clones in cultivation is not large, but following the plan adopted in the case of *T. canadensis* this epithet is here used as a group name.

ARGENTEA GROUP.

Tsuga mertensiana f. ´**argentea** (Beissner) Rehder, Bibliography. 18. 1949; *Tsuga pattoniana argentea* Beissner, Eintheil. Conif. Benennung 26. 1887 Nom., Handbuch Conif. 410. 1891.

Plants differing from the normal in the colouring of the foliage, being glaucous, grey or silvery-grey.

This is a difficult group to define since seedlings with more or less glaucous or grey foliage turn up in the seed-beds. The following names have been given to selected clones, but (as in the case of *Picea pungens* and other species where glaucous seedlings occur) intermediate shades occur, so a clonal name should not be used for plants raised from seed, nor in any case unless propagated from authentic clonal material. But see 'Glauca', below.

'Argentea'   'Blue Cloud'   'Blue Star'   'Glauca'

- - **'Argentea'** Although Beissner wrote that "blue-white or silvery-grey" forms can sometimes be found in seed-beds, he mentions a particular tree 4 meters high, so this epithet should be regarded as a cultivar name. I have seen a plant received from the nursery of Hillier & Sons, Winchester, England and there is another plant at the Arnold Arboretum (Accession No. 130.72) from the same origin. Because of its European origin it is reasonable to identify this clone with Beissner's plant. Apart from the silvery colouring to the leaves, these are more widely spaced and spreading than is normal in this species.
- - **'Argenteovarieta'** Hort ex Beissn. Handbuch Nadel. ed. 2. 94. 1909. See *Tsuga heterophylla* 'Argenteo variegata Silva Tarouca, Nadel, ed. 2. 302. 1923 (and ed. 1, ?)
- - **'Blue Cloud'** Poulsen ex Krüssmann, Nadel. ed. 3, 255. 1979. A compact form with intensely blue leaves, the colour of *Cedrus atlantica* 'Glauca'. Raised from Canadian seed and selected by D. T. Poulsen, nurseryman of Denmark.
- - **'Blue Star'** Grootendorst, Dendroflora **2**: 49. 1965. This was described as a blue-needled form. I have no further information.
- - **'Cascade'** New Cultivar. A slow-growing, very compact plant resembling Dwarf Alberta Spruce, discovered at the 4000 foot level in the Cascade Mountains in the headwaters of the North Umpqua River, Douglas County, Oregon. In the spring of 1973 the plant was 12 feet high with a trunk diameter of 8 inches at the base. The annual growth is 1½-2 inches and the leaves are shorter and more densely set than in typical *T. mertensiana*. The discoverer, J. D. Vertrees of Oregon, sent propagating material to Dr. Robert Ticknor of the North Willamette Experiment Station, Aurora, Oregon; Don Smith of Morris Plains, New Jersey and the U.S. National Arboretum. The above information was supplied by Don Smith and Mr. Vertrees.
- - **'Columnaris'** Hort. Sweden ex Krüssmann, Nadel. 321. 1955. A compact, columnar form with short branches growing upward at a steep angle; named from a plant in the Arboretum at Drafle, Sweden. This tree was 30 feet high in 1943. I cannot trace this clone in cultivation in the U.S.
- - **'Elizabeth'** New cultivar. A spreading plant, about twice as broad as high. According to a letter from Harold Epstein, January 20, 1972, the original plant was collected on Mt. Rainier about 1940 by Mrs. Elsie Frye's daughter, Elizabeth.

Mr. James Caperci of the Rainier Mountain Nursery, Seattle, Washington, claims he has the original plant, 2 feet high and 4 feet wide in 1971. George Schenk of the Wild Garden, Kirkland, Washington, has one 15 inches high and 5 feet wide in 1972. Don

Smith, Watnong Nursery, Morris Plain, New Jersey, supplied some of this information. 'Elizabeth' is the only known spreading variant of Mountain hemlock. A small plant was observed in the collection of Joel Spingarn, Baldwin, Long Island, New York, which was 6 inches high and 12 inches wide in 1973. It is a noteworthy plant.

- - **'Glauca'** Hesse, Nurs. Cat. 51:33. 1950 (as *T. mertensiana glauca*). Described as dwarf, very slow growing, leaves radial as in normal Mountain hemlock, distinctly glaucous. The rate of growth is actually moderate rather than "very slow".

The author is not sure of being able to discriminate between the color shades intended by the terms 'Argentea' and 'Glauca'. 'Argentea' may be closer to a silvery-grey.

- - **'Macrophylla'** Hort. ex Beissner, Handbuch Nadel. 402. 1891. See *Tsuga canadensis* 'Macrophylla'.

- - **'Mount Arrowsmith'** New cultivar. A seedling found in the wild by Ed. Lohbrunner of Victoria, B.C., Canada. It makes a slow-growing, conical plant with compact but irregular growth and small leaves, darker green than normal to the species. A 30-year-old plant at Mitsch Nurseries, Aurora, Oregon, U.S.A. grown under lath shade is 8 feet high and 6 feet across. John Mitsch writes that it is difficult to root from cuttings.

- - **'Mount Hood'** New name. ('Sherwood Compacta' Hort.) This is a distinctive dwarf hemlock of unusual beauty. Robert Snodgrass of the Sherwood Nursery Company, Portland, Oregon, stated in a letter of February 7, 1972, "This is an extremely slow-growing, dense, irregular, very attractive hemlock. It was originally gathered as a native plant and is much slower growing and more compact than *T. mertensiana*. Grafted plants are about 2 feet high in fifteen years". A name in Latin-form not now being acceptable, and its anglicised form having already been used for a cultivar of *T. canadensis,* a new name has had to be found. Since this is presumably the same clone that Andrew W. Sherwood reported finding at timberline on Mt. Hood (Heml. Arb. Bull. 17, 1943)—original plant was 5 feet high and the same in width—this name is suggested. Propagation has been limited.

- - **'Murthly Castle'** New cultivar. This, remarkably enough in a species so little prone to variation, must have occurred as a seedling mutation in the first introduction of this species into Britain by A. Murray in 1854, since the mother plant is one of a group of normally branched trees planted at Murthly Castle, Perth, in Scotland in 1862.

The late F. R. S. Balfour (Report Second Conif. Conference 204, 1933) recorded that it was 21 m high in 1932 and that the weeping characteristic does not reproduce from seed. There is a photograph (Fig 61.) in the same Report.

In the Report of the Third Conifer Conference held by the Royal Horticultural Society, Alan Mitchell (*Conifers in the British Isles,* 274, 1972) records its height as being

T. mertensiana 'Elizabeth'. Collection of Joel Spingarn, Baldwin, Long Island, New York, in 1973.

T. mertensiana 'Glauca'. Christy Park, Nassau County, Long Island, New York, in 1974.

28.5 m in 1970. He tells me that it is an extremely beautiful tree.

- - f. **nana** Noble, Journ. Calif. Hort. Soc. **12**: 160. 1951. A local population found by the late James Noble of California in the vicinity of Echo Lake and Summit in the wilderness area to the north of the lake.

- - **'Nana'** Welch (Dwarf Conif. 323, 1966) introduced 'Nana' as a cultivar name and quotes Mr. Gotelli's description "It does not develop a leader but has many radiating branches and glaucous foliage". Mr. Gotelli made this description from a 12″ high plant he preserved at that time, the only one he had ever seen.

I am unable to say whether or not this was a clone of the foregoing validly published but otherwise unrecorded botanical form, but I would regard it as probable, in which case the legitimacy of the cultivar name is only of academic importance. This plant is not now in the Gotelli collection in the U.S. National Arboretum at Washington, D.C.

- - **'Quartz Mountain'** New cultivar. A very slow-growing pyramidal variety collected about 1960 on Quartz Mountain by George Schenk of the Wild Garden, Kirkland, Washington.

- - 'Sherwood Compacta' Hort. See 'Mount Hood'.

- - **'Silvery'** New name. (*T. mertensiana argenteo-variegata* Hort. ex Krüssman, Nadel. 321. 1955.) *Argenteo-variegata* cannot be used as a cultivar name under *T. mertensiana* because the epithet has already been applied (in 1913) to *T. heterophylla*. Branch tips whitish, otherwise like the species. This may no longer be in cultivation.

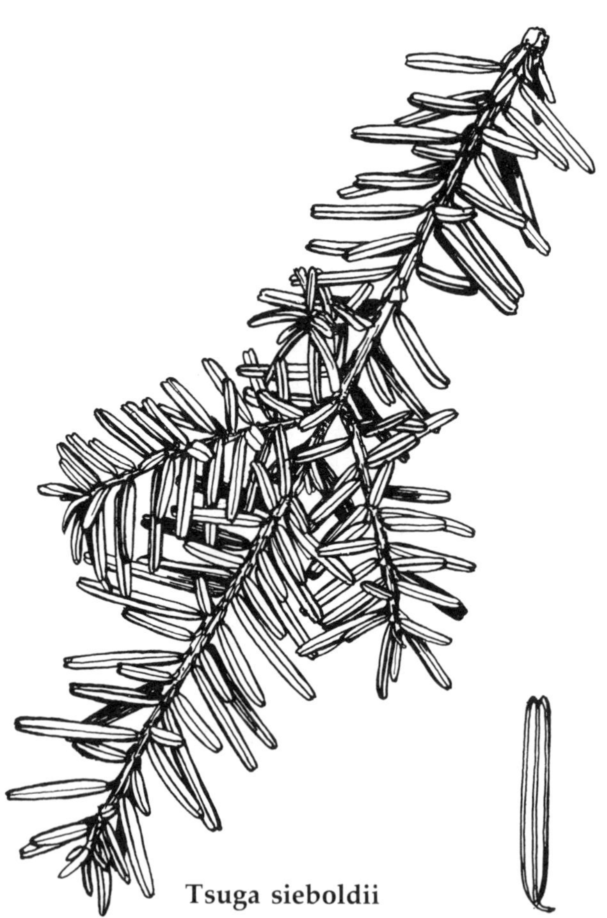

Tsuga sieboldii

Foliage of *Tsuga sieboldii*, Southern Japanese hemlock, a species in which the length of the leaf and the brightness of the white bands on their undersides (right × 1⅓) vary much, but they are always longer and less white than in *T. diversifolia*.

TSUGA SIEBOLDII CARRIÈRE, SOUTHERN JAPANESE HEMLOCK.

Tsuga sieboldii Carrière, Traité Conif. 186. 1855.
Abies Araragi Siebold in Verh. Bat. Genoot. Kunst. Wetensch. **12**: 12 (Syn. Pl. Oec. Jap.). 1830, nom.
Abies Tsuga Siebold & Zuccarini, Fl. Jap. 2: 14, t. 106. 1842.
Pinus Araragi Siebold ex Endlicher, Syn. Conif. 83. 1847, pro syn.
Pinus Tsuga Murr. ex Antoine, Conif. 23, t. 32, fig. 2. 1857.
Micropeuce Sieboldii (Carr.) Spach ex Gordon, Pinet. Suppl. 13. 1862, pro syn.
Tsuga Tsuja A. Murray in Proc. Hort. Soc. Lond. 2: 508, fig. 141-153. 1862.
Abies (Tsuga) Tsuja A. Murray, Pines Firs Jap. 84, fig. 159-171. 1863.
Picea Sieboldii (Carr.) Bertrand in Ann. Sci. Nat. Bot. sér. 5, **20**: 89. 1874.
Pinus Sieboldii (Carr.) W. R. McNab in Proc. Roy. Irish Acad. ser. 2, **2**: 213. 1876.
Tsuga Araragi (Sieb.) Koehne, Deutsche Dendr. 10. 1893.
Pinus araragi (Sieb.) Voss in Mitt. Deutsch. Dendr. Ges. **1907** (16): 63. 1908.

A pyramidal tree to 30 m in nature. Usually much smaller in cultivation.

South Honshu and the southern islands of Japan. Introduced in 1861. Uncommon in Britain. In some collections and many large gardens; more common than *T. diversifolia.*

BARK. Dark grey and smooth at first with regular horizontal folds; later cracking vertically, then into pink-grey square scales.
CROWN. Usually multiple stems from the ground and broadly conic but well defined acute apex; more open and pointed than *T. diversifolia.*
FOLIAGE. Shoot shining, glabrous pale buff; pulvini orange-red; second year shoot dull brown.
BUD. Surrounded by small leaves but more visible than in *T. diversifolia,* ovoid, narrow at base; dark orange, convex scales.
LEAVES. Irregularly spreading below shoot; of varied lengths and pectinate above, 0.7-2 cm long 2 mm broad, less regularly arranged but often more dense than *T. diversifolia,* shining dark green and grooved above, rounded and notched at the apex; two broad white bands beneath, less bright than in *T. diversifolia* and sometimes pale.
CONE. Pendulous, 2.3 × 1.3 cm, blunt ovoid-conic; scales flat-topped, very dark brown.
GROWTH. A taller growing tree than *T. diversifolia* and slightly more rapid in growth, but exact data are few.
RECOGNITION. Very similar to *T. diversifolia,* differing in the colour of the glabrous shoot, usually narrower leaves less regularly arranged and less white beneath; also in a more open and taller crown.

Bean (*Trees and Shrubs* 608. 1914, and later editions) is enthusiastic in his description of this hemlock. He says, "The grace and beauty of Siebold's hemlock makes it well worth cultivation." Wyman (*American Nurseryman* **112**(11): 49. 1960) is not so flattering. He lists this species under "Not sufficiently different as an ornamental for recommended types." The writer is inclined to agree, especially since here in the eastern states *T. diversifolia* is usually superior in health and character.

Only two cultivars in this species have been recorded.

- - **'Compacta'** Anon in Bull. Fed. Soc. Hort. Belgique. Suppl. (1883) 1887. Name only. There is no way of identifying this clone, so the name should not be applied to living plants.
- - **'National'** New cultivar. This was named from a picturesque, fast-growing tree (Accession No. 20655) in the U.S. National Arboretum, Washington, D.C. by Robert F. Doren, then curator of the Gotelli Dwarf Conifer collection. It has no definite leader. The leaves are longer and a darker green than is normal to this species, and the stomatic lines on the underside of the leaves are more blue. In 1962 this tree was less than 2 feet high; now in 1982 it is 45 feet high. This plant needs watching—it sounds to me nothing more than a tree enjoying nutritional advantages of some sort.

TSUGA YUNNANENSIS (FRANCH.) PRITZEL.
YUNNAN HEMLOCK

Tsuga yunnanensis (Franch.) Pritzel in Bot. Jahrb. **29**: 217 (1900).—Masters in Jour. Linn. Soc. Lond. Bot. **26**: 556. 1902; in Gard. Chron. ser. 3, **39**: 236, fig. 93. 1906, p.p.
Abies yunnanensis Franchet in Jour. de Bot. **13**: 258. 1899.
Abies dumosa var. *chinensis* Franchet, l. c. 258. 1899, quoad specim. Delavay.
Tsuga Brunoniana var. *chinensis* (Franch.) Masters in Jour. Linn. Soc. Lond. Bot. **26**: 556. 1902, p.p.

A small, narrow tree with level branches and foliage similar to *T. dumosa* but less pendulous and often less bright white beneath. Yunnan hemlock, Yunnan and West Szechuan, Sichuan Province, China. Introduced in 1908. Very rare; in a few collections in South England.

FOLIAGE. Shoot rather stout, pale brown above; cream-white beneath, with patches of dense, curly pubescence.
BUD. Pale brown, 2 mm shiny and lumpy.
LEAVES. Rather sparse, more or less pectinate, of varying length 1-2 cm, 2-3 mm broad, blunt or slightly acute; pale shining green above; two broad white bands sometimes very white beneath.
CONE. Nodding, 2.3 × 1.2 cm, green shiny scales with pale edges, broadly rounded; ovoid-cylindric.

Tsuga yunnanensis

Foliage and cone of *Tsuga yunnanensis*, a rare Chinese hemlock with broad and bright white bands on the under surface of the leaf. (Detail × 1⅓).

7 *PROPAGATION*

In a monographic work of this kind, treatment of the subject of this Chapter "in depth" would be out of place. The interested reader is advised to consult one or more of the many good text-books on plant propagation that are available, especially if what is in mind is any commercial application of the craft.

PROPAGATION FROM SEED

The method of producing new individuals to which Nature is almost completely restricted is the production, distribution and germination of seed. As this must be a rather haphazard and hazardous chain of events her usual method is to be prodigal in production and ingenious (sometimes extremely so) in distribution. Germination is left to the ability of each individual seed to survive, although it often has built-in mechanisms that, by controlling dormancy and its reaction to its surroundings, enable it to choose a time to germinate when, hopefully, suitable temperatures, humidity and other critical circumstances are on the way. Its immediate surroundings, whether it has fallen "by the way-side . . . upon stony places . . . among thorns . . . or in good ground" is beyond its control. But usually the situation near the mother-plant will be favourable; this is one reason why regeneration seems to take place under forest conditions so easily.

When Man elects to use this method (as of course he must do when he wants new plants in large numbers) he must circumvent Nature at the distribution stage by collecting the seed as it ripens on the tree and provide his seed-crop with a suitable environment in which to germinate. In this task Nature can teach him a great deal. He does well to listen!

The importance of seed testing and selecting seed only from stands of approved trees for such a long-term crop has long been appreciated and many countries have controlling legislation. In the United States the *International Rules for Seed Testing* (Proceedings of the International Seed Testing Association **31**(1): 1-152. 1966) are widely used. In Britain the Seeds Act 1920 and Regulations made thereunder control the purchase and sale of seeds of all the main conifer and hardwood timber trees, including the Western hemlock, *T. heterophylla*. The control does not apply to small sowings not intended for afforestation activities, but for operations on even the smallest scale the manuals available to forestry nurseries (in United States the publications of the Forest Service of U.S. Department of Agriculture such as *"Seeds of Woody Plants in the United States"* (Agricultural Handbook No. 450) and in Britain, Forestry Commission Booklet No. 43, *"Nursery Practices,"* H.M.S.O.) offer invaluable advice.

Hemlocks produce seed erratically, a good crop being produced only every third or fourth year. The cones should be gathered when they are ripe but before they open and

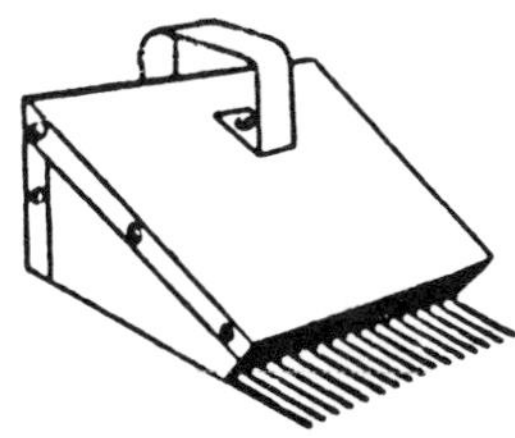

scatter the seed. According to the climate this can be in August or September, or even later. Cones from the crown of the tree are considered best but these are only reached at an increased risk of falling off the ladder and breaking one's neck. This is to be avoided, if possible. Since, except for the Mountain hemlock *T. mertensiana*, the cones are very small, collecting them can be a tedious operation. The sketch is reproduced from *Hemlock Arboretum Bulletin* 61. 1948. No dimensions or constructional details were given but it appears to be a kind of comb to be drawn downwards through the branchlets with a receptacle for the cones to collect in. It would repay a little experimentation—the spacing of the tines would seem to be critical. To release the seed the cones need to be dried in shallow trays in a spot where warm air circulates. (Anywhere uncomfortably warm is too hot and over-drying results in serious loss of vitality in the seeds.) The seeds fall out of the cones as they dry, but as these can go nearly 300,000 to the lb. the results can be very disheartening.

Tsuga seeds need a chilling period before they will germinate. This can be secured by sowing the fresh seed in prepared beds in the open (Sown broadcast at 50 gm per square meter.) The soil must be cool and moist, with 3 mm of forest floor (fallen needles) upon it, and the seeds are protected through the winter by a further covering of conifer needles until they begin to germinate. Alternatively the chilling can be provided by "stratifying" the seed in moist sand (and a coating of red lead to discourage mice) for 10 weeks at or only slightly above freezing before sowing into seed trays in April-May (at 15 gm to the tray.) In either case, success seems very largely to depend upon a close simulation of the "forest floor" conditions that favour natural regeneration. An acid soil, never any drying out, no extremes in temperature and fairly heavy shading during hot weather all seem to be important. The seedlings are left until the second autumn before being lifted. Those that come through the first summer will stand winter cold unless loosened by "frost-heave".

The young plants grow extremely slowly for several years.

CUTTINGS

At the time that the author began his studies on the variations of the Canadian hemlock the value of the recently discovered plant hormones on the rooting of cuttings was only just being realised. Since the disadvantages of grafting, the only technique available at the time, was a great hindrance to the development as garden plants of many of the plants he was studying, the author took a great interest in the new technique. His own early experiments convinced him that hemlock cuttings responded readily, and convinced himself that this was indeed a break-through he took every opportunity to interest nurserymen and collectors. The early issues of the *Hemlock Arboretum Bulletin* abound in references to their conversion, one after another, to the advantages of the new technique.

Whilst the author feels gratified at having had a part in spreading this particular gospel, the rooting of cuttings in this way has now become such universal practice that no useful service would be served by repeating these accounts here. Suffice it to say, in a book of this kind, that with the *proper* use of these hormones, *Tsuga* is now regarded in the nursery trade as one of the easiest of all conifers to root successfully from cuttings. The date of taking the cuttings does not seem to be at all critical, success being recorded from cuttings taken in Summer, Autumn and early Spring. A fairly high level of auxin application is apparently of advantage (2 or 3% of the active ingredient in a talc mix and similarly a higher-than-average treatment in the case of a liquid quick dip.) Hemlock cuttings, like the mother plants from which they came, dislike a lot of heat, like a lot of water and need carefully applied shade. Otherwise, the treatment can follow that to be found in any good text-book on propagation.

Mr. and Mrs. Don Smith of Watnong Nursery, Morris Plains, New Jersey, have recorded considerable success in the use of the Nearing frame. (*Green Scene.* 26. 1982.) The Nearing frame is a form of coldframe invented and patented by Guy Nearing 60 or so years ago. The patent specification was a north-facing, glass-topped pit which had a three-sided shed enclosure built over it and painted white on the inside, and all to precise measurements. The patent has long since expired and the name is now loosely used for any form of cold-frame sunk in the ground and provided with some form of hood, painted white on the lower side, which prevents sunlight ever reaching the glass but allows the maximum amount of other light. Don writes (29th May 1983):

> I have rooted a few hundred dwarf hemlock cuttings in my Nearing Frame each year for the past ten or fifteen years, with from 80 to almost 100% take. The frame is like many others, except that it is 20 inches deep. People who have tried it in a shallow frame have not been successful.
>
> I take the cuttings in late June or early July—when the first early growth has matured and darkened, and before the second annual growth has begun.
>
> It pays to keep the cold frame glass cover clean and tight. Covering the glass in mid-winter seems to make little difference.
>
> Water every week in mid-summer to every 2 or 3 weeks in late fall.
>
> A few cuttings may root by November, but almost all root well by May, which is when I pot them in quarts and sink the pots in sawdust in the shade. I do not move them to the greenhouse at all.

But the following letter sent to me by Dirk van Heiningen of Dover, Pennsylvania shows the danger of dogmatism in plant propagation:

> Professor Harvey Gray at Farmingdale, Long Island found that the propagation of Hemlocks from cuttings was facilitated by sealing them, hermetically, within an enclosure of polyethelene film on a greenhouse bench with a controlled bottom heat-temperature between 72 and 68 degrees Fahrenheit. The medium—⅓ sand; ⅓ screened peat moss; ⅓ ground styrafoam—is layered over the heat source to a depth of 3 inches. Cuttings are prepared long or short; then dipped ¼ inch in a 2% indolebutyric acid powder or solution and inserted in the medium about two inches in the conventional way in early November (ideally). Because the cuttings will be inaccessible for at least 10 weeks, the medium should be slightly overwatered before sealing takes place.
>
> After sealing the unit completely in plastic—the top or cover plastic should be between 8 and 18 inches above the cuttings—the heat should be immediately introduced. The enclosure can be opened after a period of 10 weeks at the end of which time, rooting will have occurred. During the rooting period, shade may be necessary on sunny days to prevent the air within the plastic house from going above 90 degrees Fahrenheit.

It is probably more difficult to carry along rooted hemlock cuttings than to root them in the first place. The writer has had some experience and offers the following alternative recommendations. Each worker must establish experimentally the method that suits him best.

The rooted cuttings should be potted and kept in a cool greenhouse or storage house until the frost period is past. Then they can be sunk in peat in a protected frame. Provide shade with snow fencing. About May 15, if the frame is in full sun, add 40 percent cloth shade about 3 feet above the lath shade. Remove the cloth shade in late September, leaving only the lath shading. When the plants are set in beds, they should again be double shaded for that season, then the next season shaded with a single shade.

TISSUE CULTURE

Although probably outside the scope of most readers of this book, tissue culture has recently been shewn to be a promising alternative to vegetative propagation of Western hemlock and other conifer genera. Research at the Oregon Graduate Center, directed by Tsai-Ying Cheng and supported by a grant from the Weyerhaeuser Company has shewn that plantlets of Western hemlock can be produced with relative ease by this method, (Cheng, T-Y. *Tissue Culture Techniques in Tree Improvement.* I.F.A. Tree Improvement Newsletter No. 28. 1976.)

GRAFTING

The value of plant hormones has been proved to be so great with the hemlocks that propagation of the garden forms from cuttings is now almost universal practice, so any discussion of grafting would seem archaic. But since there may be on occasion the need to fall back on this method, a short account must be brought in here. (It may be stated that grafting *Tsuga* in no way differs from other plants, so fuller information can be found in any good textbook on plant propagation.)

The principle of grafting is to attach a small piece cut from the plant to be multiplied (the scion) to a wound made on a growing plant with a root-system of its own (the under-stock) in such a way that the two wounds are encouraged to heal together as a single wound. If the under-stock rejects the scion and starts healing on its own, the scion must eventually die and the graft will have "failed". If the two wounds unite the graft has been a success; in due course the unwanted parts of the under-stock above the graft are cut away, leaving the new plant to develop on the root-stock provided for it.

The traditional method is to use, as under-stocks, seedlings of the same or a closely related species (*T. canadensis* was often used for any hemlock species) established in pots since the previous Spring. The actual grafting was done in the greenhouse, using a side graft in early Spring, as soon as the appearance of new "white roots" indicated movement of sap in the under-stock.

It is important to cut the scion and understock to make a nearly perfect fit. The union is firmly bound with raffia or waxed string or more usually nowadays with budding strips. After three weeks in the case, the grafts should be worked over by cutting back the top of the stock about halfway if they show callus, and remoistening the peat if necessary and laying them back for another three weeks. Eventually, when the new growth is vigorous, the young plants are moved into sheltered conditions outside. Variations in the technique of coaxing the plants to normal health and vigour are as many as the number of successful propagators. The consensus of opinion seems to be that *Tsuga* dislike heat, like plenty of water and need a certain amount of shade.

The first and most obvious reason for the general switch over to cuttings from grafting is the much greater cost of the older method. Furthermore, the grafting of hemlock was not always as successful as desired. In order to obtain a high percentage of "take", the propagator must carefully compromise between too much light and not enough light. If the "take" is satisfactory, the nurseryman can still lose as much as 30 percent in the beds and later in the nursery rows. In most cases, this is due to the imperfect healing in the graft union, leading to inadequate transmission of nutrients up and down the union. If the tissues of the stock and scion have not sufficiently inter-mingled in healing, the top of the scion may break off at the point of the union from the pressure of ice, snow, or just blowing wind. An imperfect or inadequate union can usually be blamed on poor technique, but the little understood phenomenon known as "delayed incompatibility" may also be involved.

No serious study of this problem seems to have been made. One theory advanced by some operators to explain this irritating phenomenon is based on the early foresters' belief in the existence of two varieties of hemlock. (See p. 44.) Many of the original forest trees were "white" hemlocks (and, in consequence, many present-day cultivars the same) but because they produced superior timber these trees were chosen for cutting and are now rare in the wild, whereas many red trees remain and "red" seedlings predominate. According to this theory there exists some degree of incom-patibility between these two types which shews up when "white" scions are put onto "red" understocks. This explanation contains so many imponderables, including the apparent haphazard appearance of the trouble, that until some controlled experiments are available in its support, it must be regarded as conjectural.

The reproduction of hemlocks by grafting has another mark against it. There is evidence of stimulation of extreme dwarfs by grafting them on typical seedlings of

Canada hemlock. It has been suggested that dwarf cultivars be grafted on a semi-dwarf that roots easily from cuttings. Such a one would be 'Jervis'. The writer has observed other effects on two different cultivars, 'Bennett' and 'Laurie'. In 1938 and again in 1945 the writer observed the original 'Bennett' in the garden of Ralph Lott, Eatontown, New Jersey, and also plants from both grafts and cuttings in Wittenberg Nurseries, Long Branch, New Jersey. Typical plants have horizontally spreading branches and are nearly twice as broad as high. The grafted plants have grown nearly twice as fast as the cuttings, the branches are not as horizontal, the plants are taller for their breadth, and the texture of the foliage is a little coarser.

The 'Laurie' hemlock was likewise observed in the same years, comparing the original and plants grown from grafts and cuttings. The differences were even more striking than with 'Bennett'. The habit of growth of the grafted plants was different. Although globose and multi-stemmed like the parent, the lower lateral branches were drooping and the upper lateral branchlets definitely ascending instead of being more or less horizontal throughout as in the parent and in the cutting-grown plants. The effect was much less pleasing. The contrast was further emphasized by more crowded and shorter leaves on the grafted plant. By actual count there were about 25 percent more leaves per centimeters of branchlet on the grafted plants. These considerations, coupled with the generally unsatisfactory experience with grafts, definitely favours the practice of propagating by cuttings as opposed to grafting.

Both delayed incompatibility and the effect of rootstock vigour on the growth of the diminutive forms can be overcome by techniques to encourage scion-rooting. Richard Bush of Canby, Oregon has kindly supplied me with details of what he styles the "Duplex" method. In this system the normal practice of side-grafting onto a seedling in a pot is followed. A rubber-strip tie is used, but is carefully applied with fewer twists than usual, spaced well apart so that the surface of the graft is exposed between the twists. This area is then painted over with a hormone rooting solution. Richard Bush uses a solution of 1000 ppm IBA and 500 ppm NAA in water, but I would not suppose the formula would be critical. The understock below the graft but above the first root, where it will exert a constricting effect as the rootstock swells, is then engirdled with a thin, soft metal wire, tightly twisted to hold itself in place. Finally, the whole plant is transferred to a much deeper pot which is then filled to cover the entire graft with a suitable rooting medium which will encourage rooting from the scion. The same effect would presumably be secured by plunging the pots in a bench beneath a 3-4 inch layer of the medium.

The subsequent treatment follows normal practice, the rootstock foliage being reduced in three or four stages, as indicated by the growth of the scion. One exceptional circumstance needing care is that the normal breakdown of the rubber strip in sunlight does not occur so a careful watch must be kept and the strips removed before they cause girdling of the grafts as they swell.

The progressive constriction of the rootstock by the metal wire is usually sufficient to kill the original root system in two years, by which time (if care has been taken to watch against roots emerging from the understock *above* the girdling) the plant should be "on its own roots".

8 CULTIVATION, PESTS AND DISEASES OF HEMLOCKS

As in the case of all plants, consideration of the conditions in which hemlocks thrive in nature gives a good indication of those they require in cultivation.

Hemlocks are found in abundance on hillsides in areas of good rainfall where the soil is never waterlogged in winter nor dried out in summer. It thrives in pure atmosphere and succeeds best in rather open woodland where, especially during early life, each plant enjoys partial shade, growing up into full sunlight as it matures. So in cultivation, hemlocks will not be a success in a city atmosphere, in a hot, dry, built-in spot, nor in unsuitable soil. Hemlock has a good fibrous root system and therefore lends itself to transplanting, but it does not grow well in heavy, wet soil especially if not well drained. White pine, *Chamaecyparis,* or *Arborvitae* will do better in this type of soil. In such an environment hemlocks will either die slowly or the foliage turn a sickly yellow-green and the growth be unsatisfactory. If this occurs the foliage color can be restored by applying fertilizer. The nurseryman uses the most economical method, going along the rows giving two or three handfulls of 10-10-10 fertilizer to each plant in February. The home gardener may use any good nitrogenous fertilizer, remembering that "little and often" is the safe rule. The author has a 7-foot-high hemlock hedge, one end of which is very close to a large rock maple. By fertilizing the plants near the maple twice as often and twice as much as the others, the hedge has maintained almost a uniform appearance for a period of twenty years.

When planting hemlocks, especially in the case of large plants, it is advantageous to retain as much original soil as possible and to use humus such as leafmold, peat, or compost in the backfill. Humus encourages the development of micorrhiza, which play an important role in the nutrition of hemlock. If the soil is heavy but not excessively wet, raised beds or planting the ball half on top of the ground, covering the rest with a humus and soil mix to form a low mound, is recommended.

Planting and cultural techniques for hemlock follow established good practices. Planting must be firm, and staking or guying of all but the smallest plants is essential. (More planted trees fail by rocking in the ground than from all other causes put together.) Watering, to maintain the species' high requirement for soil moisture, is important during dry weather for the first 2 or 3 years. The tolerance of hemlock to shade should be made use of (to the plant's benefit whilst overcoming its recent shock) by erecting, in important cases, a temporary screen of burlap on timber supports.

Growers of hemlock for ornamental purposes used formerly to rely mainly on small plants collected in the wild. But when these are set out in nursery rows the percentage of survival is unsatisfactory, usually 50%, often much less; they may have been dug or pulled out of dry ground, have poor root systems, or have been allowed to dry out or freeze before shipping. The collected seedlings, usually 6-10 in. high plants, are set in shaded beds in the spring or early fall in a mixture of peat or leafmold and suitable topsoil. They may be grown here for two years, then set in the field with a loose ball of soil; or small plants may be placed in clay or peat pots in a cold frame for one year and set out the next spring with a tree planter. Stocks commercially are now raised from seed, but if only small numbers of plants are required, the collecting of wild seedlings is still a good method. It is desirable to look for plants that have been grown in well-drained, moist situations. In the spring of the year when the soil is really moist, most can be pulled up by hand, others easily removed with a fox-hole digger. The author has several times collected seedlings on moist banks that have been cut over with sickle-bar mowers. They were trimmed as necessary and set out in a partially shaded bed on a northwest slope. The bed was in a well-drained area, lightened with humus and mulched with half-rotted leaves or wood chips. After two years they were set in the field in April or September with usually 90 percent survival. It is becoming increasingly difficult to find satisfactory collecting sites where one is permitted to dig, and therefore nursery grown seedlings are much to be preferred.

HEMLOCKS AND SHADE

Hemlocks are exceptionally tolerant of shade, so they are frequently recommended for planting in shady places. From this, many people have come to believe that they require, or at least prefer, shade. But this is not true—tolerance and preference are not the same thing. In full shade hemlocks grow slowly and tend to lose the lower branches, especially if the strong lateral branches are not clipped back to give the lowermost branches more light. The photograph on page 16, sent by Professor William Harlow II, illustrates the effect of shade on trunk growth. Of course, the entire plant is suppressed in heavy shade and eventually dies. The picture shows a cross section of a hemlock which grew in a mixed hemlock and hardwood stand near Syracuse, New York. It was cut in 1940 when the tree was 2 feet in diameter and two hundred years old. The core represents some seventy-five years of growth, as the closeness and number of rings indicate. When the surrounding forest which shaded the hemlock was cut away it reacted almost immediately with much wider rings, showing the power of recovery after repression.

There is, however, one important qualification which also we can learn from Nature. Seedlings in the forest are shaded during their early years, so as we might expect, young plants in cultivation (especially of the very slow-growing clones) either need some shade or will at least benefit from it until well established. Very few hemlock cultivars should be planted in deep shade, particularly where there is a considerable overhang of tree branches. Most hemlock cultivars do best in full sun, although they will get along fairly well with half sun or even one-quarter sun. Colour variations especially need at least three-quarter sun to be at their best. At least one cultivar, namely 'Cole', is best in one-quarter sun. Experience dictates that at least in this area (the neighbourhood of Philadelphia, Pennsylvania) it will not establish satisfactorily in full sun.

ENVIRONMENTAL HAZARDS

Hemlocks, like all other trees are liable to be damaged by snow and ice. Fortunately the branches on typical hemlock are nearly horizontal, hence they bend easily so any snow and ice accumulation usually slides off before the branches break. Although the wood is brittle, double-trunk trees seldom split in snow, wind, and ice storms. A few landscape architects prefer double-trunk hemlocks. The author visited a property in Chestnut Hill, Philadelphia, and counted eleven medium-sized hemlocks, all with two trunks. Some cultivars, especially the spreading ones, when young, may split quite badly from the weight of snow and ice.

Wind may cause damage in three ways, particularly in the forest. Hemlock is shallow rooted and therefore prone to windthrow; occasionally entire stands have been blown over by high winds. On record is a storm in Luzerne County, Pennsylvania, which blew down six million board feet of hemlock on one tract. Another kind of damage occurring in the forest is wind-shake, (cracks between growth rings) caused by the layers of wood separating as a result of the wind swaying and rocking the tree. These cracks make the butts unfit for lumber. A third type of damage, only important with landscape plants, is windburn on the leaves. In this latitude, Chestnut Hill, Philadelphia, Pennsylvania, the author has had experience of this with a screen of hemlock about 7 feet high where the west wind sweeping across a wide valley burned the needles year after year. This commonly occurs farther north, in northern New England and other cold windswept areas. Two cultivars, 'Stranger' and 'Pomfret', are reputed to hold up much better than the species in such exposed areas.

Hemlocks are particularly susceptible to fire damage in the forest and from uncontrolled garden bonfires, especially as young trees. Pines may withstand the heat but not hemlocks. The reason for this is that the leaves are highly flammable and the bark thin.

Although hemlocks in cultivation (in a strict sense of the word) are unlikely to suffer damage from large animals, hemlocks in the forest can be damaged or even killed by browsing of white-tailed deer. Snowshoe hares and cottontail rabbits also frequently browse on hemlock. Browsing damage of course is most serious in young trees, but porcupine are stated to gnaw the bark on larger trees. And small animals such as mice, moles, squirrels and other rodents have no respect for garden and nursery fences and will feed on seeds and seedlings wherever they are to be found.

DISEASES AND PESTS

In garden cultivation, associated with other plants and in relative isolation from large populations in the wild, hemlocks are remarkably free from diseases and pests that can reach epidemic proportions under forest conditions. So there is no need to deal in this chapter with these subjects in the detail they require and generalities are of little value. Where a need arises, the standard work, *Diseases and Pests of Ornamental Plants* by Pirone, Pascal P. (1970) will be found useful, and for those concerned with hemlock-growing on a commercial scale the Forest Service (U.S. Department of Agriculture) Manual *Silvics of Forest Trees of the United States* (Agriculture Handbook No. 271. 1965) gives a good account of the diseases and pests that have been recorded, separately for each of the American species concerned.

The Hemlock Looper is quite widespread but occurs in scattered localities in the northwest coastal area of the Pacific, the Lake states, and the northeastern states. The culprit is a pale yellow caterpillar, with a double row of small black dots along the body, which is more than 1 inch long when mature. A spray with Sevin and a miticide, when the larvae are still small, should control this pest.

The earliest pest in southeastern Pennsylvania is spider mite. Trees should be examined for infestation as early as mid-May. With extreme infestation of spider mite, the foliage turns grayish and the webs are barely visible. A spray with a miticide is recommended. Dormant-oil sprays help control this menace.

Another pest is the Hemlock Fiorinia Scale, which is prevalent only in southeastern Pennsylvania. This appeared soon after 1940 but did not become serious until the mid 1960s. It has been responsible for killing hemlocks, especially in park areas. The scales form on the underside of the leaves, causing them to turn yellow and drop prematurely. A systemic insecticide provides good control such as dimethoate (Cygon). Two sprays are recommended, one in the middle of May and one in early July.

Hemlock wooly aphid became prevalent about 1970 in southeastern Pennsylvania. This pest also attacks the two western species in Oregon. Moderate infestations cause trees to lose their vigor and heavy infestations cause premature needle drop and death. Because there are insufficient data so far, all recommendations for control are tentative. It is known that dimethoate will not control it, and that diazinon and carboryl are promising. It is recommended that a spray be applied in late September or early October to kill the adults that would winter over, then another in late June to kill the nymphs.

Gypsy Moth has long been one of our most important forest pests. Recently it has infested many wooded communities and recreation areas. It was introduced in this country in 1869 for testing the production of silk. Eggs are deposited in masses of five hundred or more, nearly 1½ inches long and ½ inch wide, on trees, stone fences, buildings, vehicles, etc. Eggs begin to hatch in the latter part of April and the hatch may last a month or more. The young caterpillars are hairy and blackish. Soon they climb to the tops of trees and drop silken threads by which they may be carried for several miles by the wind.

Second- and third-stage caterpillars are ½-1 inch long and predominantly black with orange markings down their backs.

The greatest defoliation occurs in forests where trees of the white oak group predominate. Older caterpillars may feed on hemlock, pine, and spruce. Control measures should be undertaken when the caterpillars are ½-1 inch long, although present control measures are all unsatisfactory and only partially practicable.

BIBLIOGRAPHY

The following are the principal works that have been consulted or are cited throughout the book. Periodicals are cited using the abbreviations recommended in the "Guide to the Citation of Botanical Literature" given in the *Botanical Code.* A very comprehensive list of such periodicals will be found in Rehder's Bibliography.

Arnoldia 1941 Popular Bulletin from the Arnold Arboretum, Jamaica Plain, Massachusetts.

Association of American Botanic Gardens and Arboreta. Bulletin of the *American Nurseryman, The.* 1916 Chicago, Ill.

Bailey, Liberty Hyde. 1933. *The Cultivated Conifers.* New York.

Bean, W. J. 1914. *Trees and Shrubs hardy in the British Isles.* 2 vols. London.

_______ . 1916. *ditto.* Ed. 2. London.

_______ . 1933. *ditto. Supplementary volume.* London.

_______ . 1950. *ditto.* Ed. 7, 3 vols. London.

Beissner, Ludwig H. 1884. See Jäger.

_______ . 1887 (April). *Systematische Eintheilung der Coniferen.* Erfurt.

_______ . 1887 (October). *Handbuch der Coniferen-Benennung.* Erfurt.

_______ . 1981. *Handbuch der Nadelhölzkunde.* Berlin.

_______ . 1909. *ditto.* Ed. 2. Berlin.

_______ . 1930. For Edition 3, see Fitschen.

Bongard, H. G. 1832. *Observations sur la végétation de l'ile de Sitcha.* Reprinted from *Memoires de l'Academie Imperiale des Sciences,* 6th Series, Vol. 2, pp. 119-178. 1833. St. Petersburg.

Boucher, Pierre. 1664. *Histoire veritable et naturelle de mouers et productions du pays de la nouvelle France vulgairement dite le Canada.* Paris.

Britton, N. L. and A. Brown. 1896. *Illustrated Flora of the northern United States, Canada and the British Possessions.* 3 vols. New York.

Brooklyn Botanic Gardens Record. 1912 Brooklyn, New York.

Büsgen, Moritz and Ernst Münch. 1929. *The Structure and Life of Forest Trees.* Trans. by Thomas Thomson. New York.

Carrière, Elie-Abel. 1854. *Manual des Plantes.* Vol. 4. Paris.

_______ . 1855. *Traité Générale des Conifères.* Paris.

_______ . 1867. *ditto.* Ed. 2. Paris.

Conifers in the British Isles. See Forestry Commission and Royal Horticultural Society.

Conifers in Cultivation. See Royal Horticultural Society.

Dallimore W. and A. Bruce Jackson. 1923. *A Handbook of Coniferae.* London.

_______ . 1948. ditto. Ed. 3. London.

_______ . 1966. *Handbook of Coniferae and Ginkgoaceae.* Ed. 4, revised by S. G. Harrison. London.

Del Tredici, Peter. 1983. *A Giant among the Dwarfs.*

Den Ouden, P. 1937. *Naamlijst van Coniferen.* Boskoop, Holland.

Den Ouden, P. and B. K. Boom, 1965. *Manual of Cultivated Conifers.* The Hague.

Dendroflora. 1964 Nederlandse Dendrologische Vereniging, Boskoop, Holland.

Downing, Andrew Jackson. 1857. *Rural Essays.* Edited, with a memoir of the author by G. W. Curtis and a letter to his friends by Frederika Bremer. New York.

_______ . 1875. *A Treatise on the Theory and Practice of Landscape Gardening.* Ed. 9, with two Supplements by H. W. Sargent.
Elwes, Henry J. and A. Henry. 1906-1913. *The Trees of Great Britain and Ireland.* 7 vols. Edinburgh.
Endlicher, Stephan L. 1847. *Synopsis coniferarum.*
Fitschen, J. 1930. *Beissner's Handbuch der Nadelgeholze.* Ed. 3 (See Beissner, above.)
Forestry Commission. U.K. (Published by H.M.S.O.)
    1972 *Conifers in the British Isles.* Alan C. Mitchell. Bulletin No. 33.
    Nursery Practices. Booklet No. 42.
Garden and Forest. 1888-1897. New York.
Gardeners' Chronicle. 1841 London.
Gordon, G. 1858. *The Pinetum.* London.
_______ . 1862. *ditto. Supplement.* London.
_______ . 1875. *ditto.* Ed. 2, London.
Harlow, William M. and E. S. Harrar. 1937. *Textbook of Dendrology.* New York.
Hemlock Arboretum Bulletins, Nos. 1-74. 1932-1951. Published by Charles F. Jenkins.
Hillier, Sir. Harold G. 1964. *Dwarf Conifers.* London.
_______ . 1971. *Manual of Trees and Shrubs.* (And later editions.)
Hoopes, Josiah. 1868. *The Book of Evergreens.* New York.
Hornibrook, Murray. 1923. *Dwarf and Slow-growing Conifers.* London.
_______ . 1938. *ditto.* Ed. 2. London.
International Code for Botanical Nomenclature (1982).
International Code for the Nomenclature of Cultivated Plants. 1980.
Jenkins, Charles F. See *Hemlock Arboretum Bulletins.*
Journal of Forestry. 1917 Washington.
Jäger, Hermann. 1865. *Die Ziergehölze.* Weimar.
Jäger, H. u L. H. Beissner. 1884. *ditto.* Ed. 2. Berlin.
_______ . 1889. *ditto.* Ed. 3. Berlin.
Journal of the Arnold Arboretum. 1919. Jamaica Plain, Massachusetts.
Kent, Adolphus H. 1900. *Veitch's Manual of the Coniferae.* See Veitch.
Kew Handlist of Coniferae. See Royal Botanic Gardens.
Krüssmann, Gerd. 1955. *Die Nadelgehölze.* Berlin.
_______ . 1960. *ditto* Ed. 2. Berlin.
_______ . 1972. *Handbuch der Nadelgehölze.* Berlin.
_______ . 1979. *Die Nadelgehölze.* Ed. 3. Berlin.
Little, Elbert T., Jr. See United States Department of Agriculture.
Marie-Victorin, Frère. 1926. *Notes pour servir a l'histoire de nos connaisances sur les abietacees du Quebec.* New York.
Maximowicz, C. J. 1868. *Bulletin de l'Academie Imperial des Sciences.* Series 3. **12:229.** 1868. St. Petersburg.
Michaux, Andre. 1803. *Flora Boreali-Americana.* 2 vols. Paris.
Mitchell, Alan C. See Forestry Commission, U.K.
Nelson, John P. (As Johannes Senelis) 1866. *Pinaceae, being a Handbook of the Firs and Pines.* London.
Morris Arboretum Bulletin. Published by the Morris Arboretum, 9414 Meadowbrook Avenue, Philadelphia, Pennsylvania 19118.
National Register of Big Trees. Currently maintained by The American Forestry Association, 1319 Eighteenth Street, NW Washington DC 20036.
Nicholson, Geo. 1884-1888. *The Illustrated Dictionary of Gardening.* 4 vols. and a *Supplement* issued in 1900.
Oliver, R. W. 1973. *Hedges for Canadian Gardens.* Canada Department of Agriculture Information Publication No. 889. Ottawa.
Pirone, Pascal P. 1970. *Diseases and Pests of Ornamental Plants.* New York.
Plants and Gardens. Published quarterly by Brooklyn Botanic Garden and Arboretum, 1000 Washington Avenue, Brooklyn, N.Y. 11225.

Plukenet, Leonard. 1691-1696. *Phytographia seu Stirpium illustrium et minus cognitarum icones.* 3 vols.

Rehder, Alfred. 1927. *Manual of Cultivated Trees and Shrubs Hardy in North America.* New York

_______ . 1940. *ditto.* Ed. 2.

_______ . 1949. *Bibliography of Cultivated Trees and Shrubs.*

Rothrock, Joseph T. 1880. *Catalogue of Trees and Shrubs Native and Introduced in the Horticultural Gardens adjacent to Horticultural Hall in Fairmount Park.* Philadelphia.

Royal Botanic Gardens, Kew, U.K. (Published by H.M.S.O.)

_______ . 1896. *Handlist of Coniferae grown in the Royal Gardens at Kew.*

_______ . 1903. *ditto.* Ed. 2.

_______ . 1925. *ditto.* Ed. 3.

_______ . 1938. *ditto.* Ed. 4.

_______ . 1961. *Handlist of Coniferae . . . Kew and the National Pinetum, Bedgebury.*

Royal Horticultural Society, Vincent Square, London, S.W.1.

1932. *Conifers in Cultivation.* (Report of Conifer Conference 1931. *Editor:* Chittenden, F. J.)

1972. *Conifers in the British Isles.* (Report of Conifer Conference 1970. *Editor:* Napier, Miss E.) See also Mitchell, Alan C.

Santamour, Frank S. and Harry C. Kettlewood. *The Preparation of Spruce and Hemlock Herbarium Specimens.* In "Morris Arboretum Bulletin" 14(1):15. 1963.

Sargent, Charles Sprague. 1898. *Silva of North America.* 14 Vols. Boston.

_______ . 1905. *Manual of Trees of North America.* Boston.

Schelle, E. 1909. *Die Winterharten Nadelholzer Mitteleuropas.* Stuttgart.

Schneider, Camillo. See Silva-Tarouca.

Sénéclauze, Adrien. 1868. *Les Conifères.* Paris.

Silva-Tarouca, Ernst Graf. 1913. *Unsere Freiland-Nadelholzer.* Vienna

_______ . 1923. *ditto.* Ed. 2. Vienna.

Slavin, A. D. See Royal Horticultural Society.

Sudworth, George B. See United States Department of Agriculture.

Swartley, John, C. 1938. *The Eastern Hemlock and its varieties.* In "The Arborists News." 3(4):2. 1938.

_______ . 1938. *The Eastern Hemlock and its Varieties.* In "The National Nurseryman," 46(6):2. 1938.

_______ . 1939. *Canada Hemlock and its Variations.* Unpublished Thesis submitted to Cornell University.

_______ . 1940. *The Cause and Origin of Weeping Trees.* In "Trees Magazine," 3(3):6. 1940.

_______ . 1940. *Propagating Hemlock Varieties.* In "American Nurseryman," 71(4):3. 1940.

_______ . 1942. *The Story of Red and White Hemlocks.* In "Trees Magazine," 5(4):8. 1942.

_______ . 1945. Unpublished supplement to the 1939 Thesis. (Used by den Ouden and Boom in *Manual of Ornamental Conifers,* 1965.)

_______ . 1945. *Variations and Uses of Canada Hemlock.* In "Cornell Plantations," 2(1):3. 1945. Ithaca: Cornell University.

_______ . 1946. *A Sing around the Hemlock Circle.* In "American Nurseryman," 83(7):7. and 83(8):11. 1946.

_______ . 1965. *Weeping Hemlocks* In "Plants and Gardens" 21(1):14. 1965.

United States Department of Agriculture, Washington D.C.

_______ . 1953. *Check List of Native and Naturalized Trees of the United States.* Forestry Service Miscellaneous Publication 46.

Little, Elbert T., Jr. 1971. *Atlas of United States Trees.* Vol. 1.

Forest Service Miscellaneous Publication 1146.

Sudworth, George B. *Nomenclature of the Arborescent Flora of the United States.* 1897. Division of Forestry, Bulletin 14.

_______ . 1927. *Check List of the Forest Trees of the United States.* Miscellaneous Circular 92.

von Tubeuf, C. F. 1933. *Das Problem der Hekenbesen.* In "Zeitschrift fur Pflanzenkrankheiten und Pflanzenschatz." **43**(193):242. 1933.

Webster, A. D., 1896. *Hardy Coniferous Trees.* London.

Welch, Humphrey J. 1966. *Dwarf Conifers: A Complete Guide.* London.

———. 1968. *ditto.* Ed. 2. London.

———. 1979. *Manual of Dwarf Conifers.* New York.

Wilson, Ernest H. 1916. *The Conifers and Taxads of Japan.* London.

Woody Plant Registrations. List One. 1949. American Association of Nurserymen.

Ditto. List Two. 1950.